How To Produce Effective Operations and Maintenance Manuals

Mike Tidwell

American Society of Civil Engineers
1801 Alexander Bell Drive
Reston, Virginia 20191-4400

Abstract: This guide will aid in the process of writing and managing the production of operations and maintenance (O&M) manuals. Featuring examples of standard operating procedures and other O&M manual entries, the book demonstrates the wide range of options available. It contains information on defining the content of the manual, estimating production costs, determining and meeting a production schedule, details of producing a manual, and placing an O&M manual online. This guide will provide both novice and experienced O&M manual producers with the hands-on knowledge needed to make the most of the latest techniques for producing effective O&M manuals.

Library of Congress Cataloging-in-Publication Data

Tidwell, Mike, 1956-
   How to produce effective operations and maintenance manuals / by Mike Tidwell.
     p.   cm.
   ISBN 0-7844-0011-3
    1. Communication of technical information. 2. Technology—Documentation.
   3. Technical manuals. 4. Plant maintenance—Handbooks, manuals, etc. I. Title.

T10.5. T53 2000
808'.06665Cdc21

                        00-028892

To my son and best friend, Clay. If we only had an
O&M manual for teenagers!

Table of Contents

Foreword

In today's competitive business environment, utilities are feeling increasing pressure to become more efficient and effective. Efficiency and effectiveness result from good decision-making, beginning with strategic planning and carrying through to day-to-day operations. This book addresses one of the most fundamental tools a utility has for day-to-day decision-making: the Operations and Maintenance (O&M) Manual for its facilities.

Many utilities make a significant investment in producing their O&M manuals. Too frequently, these manuals in which so much is invested are seldom, or never, used and the investments are wasted. The author's goal is to help those who prepare O&M manuals to avoid this fate.

The author points out the common pitfalls in preparing O&M manuals and offers practical guidance on how to prepare an O&M manual that is a decision-making tool. His approach is based on the following six steps:

- **Identify the audience** – determine who are the intended users of the O&M Manual.
- **Define the needs** – analyze what will help the audience do their jobs efficiently and effectively.
- **Prioritize the needs** – define what the O&M Manual should provide the users and what should be provided elsewhere.
- **Design effective communication** – plan the level of writing and the presentation methods that will reach the audience.
- **Involve the audience** – have intended users test the manual throughout its development to make sure it will meet their needs.
- **Include an updating process** – to make sure the O&M Manual evolves as users' needs evolve.

The author begins with the fundamentals. He describes the basic skills that are required to produce an effective O&M Manual. He gives practical guidelines for organizing the manual, writing to reach the audience and incorporating graphics that help operators. He gives useful suggestions for producing an O&M Manual, from initial planning through desktop publishing.

The author illustrates the role the computer can play in producing an effective O&M Manual. He shows how computer technologies, when used correctly, can keep a manual accessible to the audience and up-to-date. Using computer technologies does not reduce the need for the fundamentals, but these technologies can transform a good manual into an outstanding one.

The examples the author presents draw primarily from his extensive experience with utilities, such as water and wastewater treatment plants. However, the book provides useful guidance for producing an O&M manual for any type of facility.

> *Of all those arts in which the wise excel,*
> *Nature's chief masterpiece is writing well.*
> - John Sheffield

Brooks W. Newbry, PhD
Director, Utility Management Consulting
Greeley and Hansen Engineers

Acknowledgements

I've often wondered why so many writing projects fall behind schedule. The simple truth is that we live in a very busy world. Most of my friends and colleagues, professional and nonprofessional, are constantly challenged by Father Time. I want to express a sincere thanks to everyone who has helped to get this book to print. Preparing a text on the effective production of O&M manuals was no easy task, especially when one considers the fast moving pace of computer technology.

A special thanks goes to my friend and colleague Brooks Newbry for his ongoing support, not to mention preparing an excellent foreword for the book. I'd also like to thank all my friends and colleagues at Greeley and Hansen Engineers for their continued support and encouragement.

Thanks to Joy Chau, Editor, American Society of Civil Engineers, for her guidance, direction, and patience throughout the development of the manuscript. Charlotte McNaughton, Manager of Book Production, American Society of Civil Engineers, was also instrumental in guiding me through to the book's fruition. Ladies, thank you so much.

Finally, thanks to my wife, Gaye, and son, Clay, who not only support me endlessly in my endeavors, but also sacrificed a portion of our time together to get this book to print.

Introduction

Many facility operators find their operations and maintenance (O&M) manuals either outdated, inaccurate, too technical, or difficult to understand—in other words: useless. Sadly, many manuals are shelved and left to gather dust while day-to-day facility operations are conducted using worn-out manufacturers' literature and makeshift operators' notes. Most will agree that an O&M manual is virtually worthless if not used and updated regularly.

O&M manuals are often viewed as collections of manufacturer and vendor product literature randomly assembled into loose-leaf binders. However, today, high-technology demands accurate, comprehensive, and more easily understood O&M manuals. Both managers and operators want their facilities to perform efficiently and economically. While not all facilities are high-tech, most require definitive manuals that cover start-up, shutdown, and emergency operations, at a minimum. Manufacturers often provide information for individual equipment items, but not for an entire facility as an integrated system. Instrumentation and control systems alone can require extensive operation, maintenance, and troubleshooting documentation. Many facilities also require detailed supplemental information about various operation and maintenance techniques, sampling procedures, laboratory procedures, etc.

So, whose job is it to write an effective O&M manual? Often, the design engineer produces O&M manuals. Preparing an O&M manual for a large, complex facility can be a tremendous task. For example, an effective manual must document all facility processes and equipment, manufacturers' equipment catalog information, and unexpected operational procedures and policies to name only a few.

Engineers and equipment designers who usually have many years of college education and technical experience often tend to provide highly technical documentation using language that can be perceived as confusing and pretentious. *A clear and concise manual benefits the staff that maintains and operates the equipment.* Information affecting the efficient operation and maintenance of the equipment must be included, but not using over-technical jargon. In other words, an O&M manual should be written at the level of the staff involved in the day-to-day operations and maintenance of the facility.

Sometimes the facility owner requires that new manuals be prepared in the same format as existing documentation without correcting any inadequacies or ambiguous information in the existing manuals. The result is a new version of the same old manual and, like its predecessors, will likely be ignored by the users.

Many privately owned manufacturing facilities and larger public utilities have the staff and the budget to produce and update their own manuals. The staff must, however, be given adequate support and flexibility when producing and updating these documents.

An O&M manual must be viewed as a constantly changing document. As operational and maintenance procedures change due to equipment upgrades or replacement, seasonal variations, and process conditions, so must the O&M manual. An up-to-date O&M manual can minimize the learning curve and mean all the difference to a newly hired staff.

Engineers and manufacturers are under pressure to provide the highest quality products possible within a defined budget. Pressure also applies to writing O&M manuals. This handbook will act as a helpful guide to those who write technical operations and maintenance manuals. The book also provides technical writing tips and suggestions to help anyone who prepares most types of technical documents. While the examples in this book apply to engineered facilities like pumping stations, water and wastewater treatment plants, power plants, etc., the information can be applied to any type of facility.

The author also recognizes that most companies have their own ideas and policies about how operations manuals should, and should not be produced. This book can only act as a guide for the manual writer. It is not intended to replace existing procedures that work well, but to help improve those that do not. Also, many state agencies have developed standards that outline content and format for facility operations manuals. Be sure to check with the appropriate governing agency to ensure compliance with current standards.

This book is divided into sections that cover the basic contents of an O&M manual, estimating and producing an O&M manual, updating O&M manuals, and placing O&M manuals online. Sections also cover technical writing and editing skills and desktop publishing.

Chapter 1

Defining the Manual's Audience, Content, and Level of Detail

Often, the technical writers assigned to prepare an O&M manual aren't technical writers at all, but members of the facility design engineering staff. While they may be good technical writers, converting technical wording from drawings and specifications into a concise understandable document can be challenging. One can easily become lost in most engineering document legalese and jargon alone. It is, therefore, very important to accurately define the manual format, content, level of detail, and audience early in the manual preparation process. This chapter helps you define these key areas.

Producing a document that will be used frequently and viewed by its users as an integral part of their facility is every O&M manual author's goal. Many technical writers, however, often find themselves bogged down by unrealistic deadlines, inadequate budgets, poorly defined scopes of work, and procedural red tape. The technical writer's job is to see that such obstacles do not mar the concept of a well-written, accurate operations manual.

The task of preparing an operations manual is often considered secondary, unimportant, and even dispensable by those not directly involved with the task. When project budgets are strained, the operations manual can be one of the first project elements to be trimmed or even eliminated. Therefore, it is imperative that the client and engineering staff clearly understand the importance of the operations manual for a new or upgraded facility.

The operations manual should mirror the operations function of any facility. While many experienced operators and maintenance technicians may comprise the plant staff, every facility has its own unique operational and maintenance needs. A well-compiled operations manual will enable inexperienced and veteran staff members alike to quickly locate information they might need about the facility. Such a manual should contain visual information such as tables, figures, drawings, etc. An effective O&M manual will be a successful mix of these and other forms of visual information. In addition, if the manual will be updated often, experience and knowledge from veteran staff can be incorporated easily into the manual.

The End-User

Manual users don't need, nor do they want, technical jargon or legalese clouding their understanding of equipment or process descriptions. What they *do* need is an accurate, concise document that gives a clear understanding of the equipment and procedures. Technology continues to produce more advanced equipment and processes every day. Clear and concise operations and maintenance manuals can help minimize new personnel orientation time, decrease downtime for repairs, and increase the overall usefulness of the equipment. Efficient operation and maintenance practices can also increase equipment life, reduce operating and repair costs, and increase process performance.

Preparing a quality manual sounds like an easy enough task, right? It depends. First, if a writer is to effectively convey ideas to the user, he or she must know the users' basic information needs:

- Who will be using the manual?
- Is the user familiar with technical manuals?
- Does the user have a high school or college education?
- How technically adept is the user?
- Can the user read and understand English?

Manual users are referred to as the *target audience*. Technical writers should focus their skills on adequately conveying the information in the manual *to these individuals*. If you don't know the answers to these and other important questions before you start preparing the manual, you will not be able to define the manual's contents. You might be able to correctly guess some of the answers, but what writer wants that kind of guesswork and gambling? It's difficult enough to convey information when the audience is well-defined. So, interview the end-users of the manual and find the answers *before* you start preparing the manual.

Effective O&M manuals should always be written for the end-user, not the management staff. Address management's manual needs separately with a supplement, or in a separate document altogether. After all, the main purpose of the manual is to make *operations and maintenance* procedures easier.

If the manual is to be written for a specific facility process or operation, indicate that intent at the beginning of the manual. This will help avoid confusing the manual users. The following paragraph is an example:

> This manual contains specific information related to process control of the City Water Treatment Plant. This information includes instructions and guidelines relating to process startup, shutdown, troubleshooting, and other process control items. Refer to the *City Water Treatment Maintenance Procedures Manual* and the *City Water Treatment Plant Electrical and Instrumentation Guide* for additional information.

If such a statement is not included in multi-volume manuals, the user might inaccurately conclude that any one volume is a complete reference for all facility operations. It also enables the users, including new employees, to easily locate supplemental information in the manual.

A second technique for identifying separate multi-volume manuals is to provide clear, definitive titles for each volume. Instead of simply entitling each volume in sequential numerical order such as Volume 1, Volume 2, Volume 3, etc., include a volume name. For example, **Volume I - General Plant Information** or **Volume 2 - Operations Information**. This approach might seem a bit overly simplistic, but it makes the first step of locating important information that much easier.

Many manuals will be written for several areas within the facility. In such cases, each section within the document should be clearly classified to provide easy-to-locate information for each of these areas. For example, instructions on how to calibrate an electronic flow meter should not be included in the same section describing the correct technique for taking a water sample. Calibrating electronic flow meters requires a completely different set of skills and tools than does sampling water.

It can be tempting to include all information about a certain subject within a single section. Again, the manual should be formatted and prepared for the most efficient use by the end-user. (See Chapter 2 for additional information on formatting and manual content.) Consider the following examples showing correct and incorrect inclusion of information within a manual section:

> Each potable water booster pump is supplied with sup-
> plementary water for pump seal lubrication. A valved tap
> on the seal water supply line can also be used as a pota-
> ble water sampling point.

While the information in the preceding paragraph may be accurate, it is a bit confusing. Several questions may arise about the seal water line:

- Should a seal water line be tampered with at all? After all, it is a potable water sampling point.
- What does sampling have to do with pump seal lubrication?

The paragraph could be better written:

> Each potable water booster pump is supplied with sup-
> plementary water for pump seal lubrication. (A valved tap
> on the seal water line can also be used as a potable water
> sampling point. Refer to Section 12 - Sampling for more
> information on potable water sampling points.)

When the second sentence of the paragraph is separated by parentheses, the relationship between the seal water and its possible use as a sampling point is clearly defined. Another approach is to not mention using the seal water tap as a sampling source in this section and include it in a more appropriate section. While the end-users may not be consciously aware of this approach of separating equipment information items, they will definitely find the manual easier to read and understand.

A common mistake writers make is to assume erroneously that the end-users are always technically skilled. Obviously, many facility operators and technicians are highly skilled. However, the manual should be written so readers possessing only limited skills can thoroughly understand the document. This doesn't mean the document should be written at the first grade level, but it should be easy to read and understand ... *the first time through.*

If you have a question as to whether your document falls into the "easy-to-read" category, try this: Ask a friend or colleague to read the section in question. If half way through the section your friend has to go back to the beginning and start over to understand it, you probably should rewrite the section. Even if your friend is unfamiliar with the subject, he or she should be able to read and comprehend the section easily the first time through.

The following is an example of a manual section that is written unclearly:

> Hydraulic overloading of the primary sedimentation unit can cause a substantial decrease in the overall equipment performance. Solids removal rates may decrease dramatically while solids flow-through will likely increase. Proper hydraulic distribution among in-service units will facilitate optimum individual unit performance. Consult the manufacturer's performance criteria.

Huh? This is the same section, rewritten to make it much easier to understand:

> Each primary sedimentation unit is designed to operate efficiently within a specific flow range. Operating any of the units at a flow rate higher than the recommended maximum may cause the unit(s) to work less efficiently. Higher flow rates can cause fewer solids to settle in the unit, thus decreasing solids removal rates. Solids that do not settle will exit the unit, thereby increasing the solids flow-through

rate. If one or more units are operating at too high a flow rate, verify that other units are receiving adequate flow. Redistributing flow to these units can often relieve the overloaded unit or units. Refer to each unit's manufacturer's catalog and operating manual for specific information regarding minimum and maximum recommended flow rates as well as typical performance information.

The revised section may be a bit longer, but it can be much more easily understood. Also, the references to information contained in the manufacturer's catalog and operating information is better defined.

Because O&M manuals are often used as training tools, the manual should be easily read and understood by entry-level employees. How skilled are entry-level personnel? Usually, this is a question the facility manager can answer. After all, minimum technical skill requirements for new operators and technicians are facility-specific issues. These minimum requirements may even vary at each facility location. In cases where the manual is to be used primarily as a training tool, you should obtain a list of minimum skill requirements and write the manual to those standards.

This does not mean that important information meant for more highly skilled technicians should be excluded altogether. It does mean, however, that much of the information may need to be divided into separate sections. This approach will ensure that entry-level persons will thoroughly understand the material, yet information regarding procedures will also be easily accessible. Following is an example paragraph for the primary sedimentation process at a typical wastewater treatment plant:

The primary treatment process provides conditions that allow solids in the wastewater either to settle to the bottom or float to the surface of a settling tank, or clarifier. This reduces the total suspended solids (TSS) and the biochemical oxygen demand (BOD), thereby reducing the loadings on the downstream-activated sludge process. The clarifiers also serve as thickeners for the primary sludge that settles, allowing gravity to compact the sludge in the bottom of the tank.

Most entry-level persons at a wastewater treatment plant are expected to be familiar with terms like TSS and BOD. The same paragraph, written for more advanced operators, might read:

> The primary treatment process provides hydraulic conditions that allow solids in the wastewater stream flow to settle to the clarifier bottom. Particle settling velocity profiles have been developed for each primary clarifier, thereby mapping each unit's hydraulic and settling characteristics. A correlation between TSS and BOD affecting settling velocities has also been determined. Table 2.3 Advanced Process Control shows these correlation values plotted over time for each....

Obviously, this paragraph delves more deeply into the technical aspects of operating a primary clarification process. For an experienced plant operator, information such as this not only may be interesting, but also a necessary, vital part of the successful day-to-day operations of a facility. On the other hand, a trainee might find it very confusing and complicated. Such detailed information should be placed in a separate section with the title "Advanced Process Control."

Determining the Manual's Content

Once you have determined your audience, your next task before compiling an O&M manual is to define its content. A manual's content includes specific types of information to be covered: text, photographs, drawings, reference tables, etc. Sometimes the facility owner provides you with format and content guidelines for the new manual. These guidelines define *what* the manual should contain, *how* the manual should be written and formatted, and *for whom* it should be written. Unfortunately, the manual's clear direction and content are rarely laid out in advance.

You may find it difficult to determine the manual's content because of the often-large number of individuals involved in the preparation and approval process. Staff from management, engineering, facility operations and maintenance, and regulatory agencies all can have requirements (or strong opinions) as to the manual's content.

Workshop Sessions

To make sure everyone's ideas are weighed fairly, you should schedule several meetings with a cross-section of end-users *before* manual preparation begins. In this way, everyone has a chance to provide input into the manual. Furthermore, these workshop sessions facilitate communication between the design engineer, manual author, and end-users. Sometimes, information gained in these sessions will completely alter the original concept of the manual format and content...often for the better.

There is no easy way to estimate the number of workshop sessions that will be required for a particular manual project. So, schedule at least one workshop session for every 50 to 100 pages of text to be written. In any case, the sessions should be enough to help you become intimately familiar with the end-users' manual needs. Although not every suggestion can be implemented, the meetings ensure that the manual's content is well-defined and the end-users will be more accepting of the process from the very beginning.

Workshops are needed even if an O&M manual is to be produced within a single company. You have clients to please, even though they may be co-workers. Be careful, though, that staff from the end-user group are not over-powered by management. Everyone should be able to freely voice his or her concerns about producing a useful O&M manual that will make everyone's job at the facility easier.

Taking staff away from their normal work duties to participate in these workshops could be a problem. You must convince management staff that time spent in these sessions will result in a useful, high-quality O&M manual. Recommend that management schedule one or two-hour sessions weekly at first, followed by longer, more frequent workshops as the need arises. You may only need to arrange one or two workshops for a manual project. On the other hand, you may need several sessions.

The following are suggestions on how to arrange, schedule, and manage successful workshop sessions:

- Arrange workshops with as few individuals as possible while still obtaining a representative cross-section of users.

- Most people agree that short, productive sessions are more beneficial than longer, regularly scheduled meetings.
- Schedule the workshops at least two weeks in advance.
- The chief writer should chair the workshop session. Be ready to make on-the-spot decisions about the manual compilation process. Meeting notes taken by the chairperson or assistant should be promptly transcribed and distributed to all workshop attendees.
- Use handwritten flip charts during the workshop sessions. Flip chart sheets completed during the session can be taped to a wall and be easily referred to. The flip charts can also be used to help document the workshop.
- Resolve workshop topics or issues as soon as possible. All workshop participants should be asked to resolve questions and issues (for which they are responsible) in a timely manner. Unresolved issues can tend to stay unresolved, thereby holding up progress.

Following are a few basic questions the writer should research early in the manual preparation process:

- Is the facility large or small?
- Is the facility private, municipal, or government-owned? How will this affect decisions about the manual?
- Can all O&M topics be covered in one manual or will multiple volumes be necessary?

The answers to such questions determine the direction of the O&M manual. For instance, if the facility is large, the manual topics may need to be divided into several volumes. The method and format for dividing up the manual will likely be unique for every project. Because O&M manuals convey a lot of information, they often can't be compacted into a single three-ring binder. If, early in the project, it becomes obvious that several volumes will be needed, inform the end-users that you plan to develop several volumes.

Don't let the prospect of multiple volumes scare you. The formula is the same for a 50-page operations manual or a 2,000-page version. It just takes more time and paper to develop the larger manual. Lack of early planning and scheduling can, however, cause the simplest manual project to grow into an unmanageable monster. To adequately define the content of any O&M manual, you must first know exactly what the end-user is expecting. Don't assume that you already know what the end-user needs, because your instincts may be wrong and the project will suffer.

Before a final outline is prepared, give the end-user the opportunity to review the manual format and breakdown. You should insist on a thorough review in preparation of the final version of the manual.

Once the end-users have reviewed the draft outline, prepare a revised outline. This revised outline will probably not be the final outline version, however. The outline will likely change as you compile the manual sections. Any major alterations should always be approved by the end-users.

Defining the Level of Detail

Your next task is to determine the necessary level of detail. Level of detail is defined as the amount and type of information provided for any particular facility, equipment item, or process. Defining the amount and type of information is a major step in manual planning and should not be taken lightly by either the end-users or the manual author.

For instance, should the equipment description sections include detailed information about every facet of the item? Should the sections include only cursory information? Or, somewhere in the middle? It may sometimes be in the end-user's best interest to provide more detail in some areas of the manual and less in others. For instance, if a facility staff focuses on preventive maintenance, the manual must reflect that in its detail.

One of the major constraints affecting a manual's level of detail is the available budget. Obviously, a manual with a high level of detail will take more time and resources to produce than one with less detail. When budgets are limited, end-users must ultimately decide on the manual's level of detail, and that decision must be made early on.

You can, however, make future revisions to the manual easier. Section and subsection titles can be included in the manual with no accompanying text. These section headers can be accompanied by a phrase like "To Be Included Later." Following is an example from a manual table of contents:

CHAPTER 4 - INSTRUMENTATION AND CONTROL SYSTEMS

4.1   Process Control Descriptions
4.2   Component List by Process
4.3   Component Calibration Procedures
4.4   Component Parts Lists (to be included later)
4.5   Process Alarms
4.6   Troubleshooting (to be included later)

Including the sections suffixed with "(to be included later)" not only informs the reader that the information is not yet available, but also that the topic has not been overlooked. These marked subsections will, in all likelihood, not be forgotten when revision time comes. These sections can even include tabs and blank pages. This allows the end-user to locate the section easily and jot down notes to be included in future revisions. Providing tabs for these sections will also allow additional pages to be easily inserted into the proper section of the manual. Merely providing blank areas in the middle of text passages will make future revision difficult. Remember that budget constraints may limit the amount of information included in the document, but that doesn't preclude efficient planning for future revisions.

Even if the budget is not a major concern, you still must define the level of detail. An O&M manual that contains page after page of unnecessary information is *not* an effective tool. An O&M manual should be thought of as a reference document, not an encyclopedia. It should be as concise and accurate as possible. If the process or equipment item in question is very sophisticated, it may be necessary to dedicate a large section to it. If so, the section should be broken up into subsections or subheadings as often as possible. This will make the document much easier to read and provide "markers" for the reader to find information easily.

As mentioned earlier, the best way to get constructive feedback from the end-users is to communicate directly with them. The value of collaboration between technical writer and end-user cannot be overstated. You must ask the end-users to point out what level of detail they need in the final product. If they aren't sure, be ready with example chapters or sections. These example chapters can be taken from other operations manuals (with the owner's permission, of course), or be purely fictitious.

Most operations manuals fall into one of three categories:

Low Level of Detail
Moderate Level of Detail
High Level of Detail.

Following are examples that can be used as guidelines.

Low Level of Detail

With a low level of detail, general facility and process descriptions are short, (100 to 500 words) and do not elaborate on related processes.

Example:

> Pump Station No. 1 is located within the fenced perimeter at the existing Thompson Reservoir site near the Mountain Ridge subdivision, (see Location Map 12.2). The two existing aboveground steel reservoirs at the site (2.5 mg and 4.0 mg) furnish water to the Thompson Pressure Zone. This pressure zone is located south of Highway 89. Pump Station No. 1 draws water from the reservoirs' 18-inch pipeline and pumps to the Smith-Jones Pressure Zone, which surrounds the reservoir site. The pump station was constructed to provide adequate system water pressure and flow for this service area.

Specific equipment items are described in general terms and detailed equipment information is left to manufacturers' literature or other supplemental documents.

Example:

> The three vertical turbine pumps located at Pump Station No.1 are powered by 40 horsepower variable speed drive motors. Variable speed motors allow the pumps to be operated over a wide range of speeds. For detailed informa-

tion on the pumps and variable speed drive units, see manufacturer's literature for each.

Process or equipment operating procedures are complete, but concise. While information necessary for proper operation is included, details are kept to a minimum. Also, abnormal and/or emergency operating procedures may even be excluded at this level of detail.

Example:

> Each of the booster pumps' adjustable speed drive (ASD) units are equipped with a three-position hand select switch labeled ON-OFF-AUTO, located on the front of each control panel. When this switch is placed in the OFF position, the pump will not operate. When this switch is placed in the ON position, the pump will start in manual mode and its speed can be controlled by the manual speed control adjustment knob located on the front of each ASD panel. When the ON-OFF-AUTO switch is placed in the AUTO position, the pump is controlled automatically by the programmable logic controller (PLC) located within the main control panel.

Manual users are generally directed entirely to manufacturers' literature regarding equipment troubleshooting at this level of manual detail.

Example:

> Pump and adjustable speed drive troubleshooting should be performed in strict accordance with the manufacturer's instructions.

Equipment information lists are generally very concise at this level of detail. Manual users are directed to the manufacturer's literature regarding specific equipment item and performance information.

Example equipment list:

Vertical Turbine Booster Pumps

Quantity   3

Manufacturer   Super Pumps, Inc.
                  2000 North St. Railroad
                  Anywhere, MA 78787
                  (555)123-4567

Supplier   Pumps, Inc.
                  P.O. Box 0001
                  Pumps, MA 92222
                  (555)765-4321

Motor Type   TEFC

Manufacturer   Electric Motors, Inc.
                  123 Super Road
                  Motors, CT 06001
                  (555) 232-2323

Supplier   Motors, Inc.
                  999 Rough Road
                  Sacramento, CA 95005
                  (555) 333-3333

Horsepower   75

Type   Variable Speed

Moderate Level of Detail

Facility and process descriptions are moderate in length (approximately 200 to 1,000 words) and provide some detailed information about specific and related equipment and processes.

Example:

> Pump Station No. 1 is located within the fenced perimeter at the existing Thompson Reservoir site near the Mountain Ridge subdivision. The two existing aboveground steel reservoirs at the site (2.5 mg and 4.0 mg) furnish water to the Thompson Pressure Zone. This pressure zone is located south of Highway 89. Pump Station No. 1 draws water from the reservoirs' 18-inch pipeline and pumps to

the Smith-Jones Pressure Zone, which surrounds the reservoir site. The pump station was constructed in 1982 to provide adequate system water pressure and flow for this service area.

Pump Station No. 1 consists of a concrete block building that houses three vertical turbine pump assemblies with room for a future fourth pump, a 16-inch bypass line, heating and ventilating equipment, and related piping, electrical, and instrumentation equipment. The building is automatically heated, cooled, and ventilated. In the event of a power failure, the bypass line will reduce surge conditions in the Smith-Jones Pressure Zone until the Hamilton Pressure Zone starts supplying water to the pressure zone. The electrical system for the pump station is also sized for the ultimate addition of a fourth vertical turbine pump and its accompanying variable frequency drive unit.

A separate room, accessible from the exterior only, was provided to facilitate the future addition of chemical injection equipment. Initially, however, there are no chemical injection facilities at the pump station.

Specific equipment items are described in moderate detail. The manual user is provided with most information necessary for adequate operation and maintenance of the equipment. The user is still directed to manufacturers' literature for additional detailed information regarding specific equipment items.

Example:

The three vertical turbine pumps located at Pump Station No. 1 are powered by 40 horsepower variable speed drive motors. Variable speed motors allow the pumps to be operated over a wide range of speeds. This allows a great degree of operations flexibility regarding pump selection and performance. Adjustable speed drive (ASD) units are provided to power each vertical turbine pump motor. These ASD units can be operated manually by the operator or automatically by a programmable logic controller (PLC). The PLC can automatically analyze each pump's performance and adjust its speed for the most efficient performance. In addition, each pump can be operated independently of the others, allowing for very flexible operating parameters. For detailed information about the pumps and variable speed drive units, refer to the ASD manufacturer's literature.

Each process or equipment operating procedure is described completely. Most operations and maintenance personnel should gain a fair understanding of the operation of this equipment or process after reading this section. However, theoretical operating procedures are rare or non-existent and abnormal and/or emergency operating procedures are brief at this level of detail.

Example:

Each of the pump's adjustable speed drive (ASD) unit is equipped with a three-position hand select switch labeled ON-OFF-AUTO, located on the front of each panel. When this switch is placed in the OFF position, the pump will not operate. When this switch is placed in the ON position, the pump will start in manual mode and its speed can be controlled by the manual speed control adjustment knob located on the front of each ASD panel. When the ON-OFF-AUTO switch is placed in the AUTO position, the pump is controlled automatically by the PLC located within the main control panel.

The PLC will turn pumps ON and OFF and control their speed based upon system demand to meet the desired system PRESSURE SET POINT. When functioning in the automatic mode, the PLC performs the function of the "pressure controller." The pressure control system is located within the main control panel. The control panel has a three-position hand selector switch labeled MASTER CONTROL LOCAL-OFF-REMOTE, a system pressure set point adjusting knob labeled PRESSURE SET POINT, a 0-100 psi pressure meter labeled DISCHARGE PRESSURE, a 0-20 psi suction pressure meter labeled SUCTION PRESSURE, a 0-6,000 gpm flow meter labeled FLOW, a reset push-button switch labeled ALARM RESET, and an alarm annunciator panel labeled ANNUNCIATOR.

Some reference to troubleshooting is usually made at the moderate level of manual detail. However, manual users are still directed to manufacturers' troubleshooting literature to obtain detailed information.

Example:

> If any of the vertical turbine pumps do not appear to be operating correctly or efficiently, the following list of troubleshooting procedures should be employed to aid in located the problem.
>
> If the pump is making abnormal noises during operation, try to determine where the noise is originating. Inadequately lubricated bearings or seals are often the cause of excessive noise. Check that adequate seal cooling water is being supplied to the pump shaft seal. Check that the motor has been adequately lubricated.
>
> If the pump's flow rate or discharge pressure is below normal, the pump suction valve could be partially closed or the suction piping partially obstructed. Check that the suction valve is adequately open.
>
> The equipment manufacturer's catalogs also contain troubleshooting information regarding most problems normally associated with their equipment. If these catalogs do not address a particular problem, call the equipment supplier or manufacturer and request this information directly.

Equipment lists generally provide adequate information at this level of detail. Manual users are directed to the manufacturer's literature regarding detailed specific equipment item and performance data.

Example equipment list:

Vertical Turbine Booster Pumps

Quantity   3

Pump Column Size	8-inch
Discharge Pressure Rating	76 psi
Available NPSH	58-65 ft
Required NPSH	13 ft
Number of Pump Stages	3
Design Capacity	1,500 gpm @ 118 ft TDH
Manufacturer	Super Pumps, Inc. 2000 North St. Railroad Anywhere, MA 78787 (555)123-4567
Supplier	Pumps, Inc. P.O. Box 0001 Pumps, MA 92222 (555)765-4321
Motor Type	TEFC
Manufacturer	Electric Motors, Inc. 123 Super Road Motors, CT 06001 (555) 232-2323
Supplier	Motors, Inc. 999 Rough Road Sacramento, CA 95005 (555) 333-3333
Horsepower	75
Type	Variable Speed
Speed (max)	1,770 rpm
Power	480v, 3 phase, 60 Hz
Frame No.	404VP
Motor Efficiency @ Rated Capacity	93%

High Level of Detail

Naturally, the term "detailed" can be very subjective. What one person may perceive as detailed may be very general to another. For purposes of example, the term means the total amount of useable information in its collective form. Detailed operations manuals are those that include large amounts of pertinent information regarding most of the facility or equipment.

Facility and process descriptions are detailed (approximately 500 to 2,500 words) and provide detailed information about specific and related equipment and processes.

Example:

> Pump Station No. 1 is located within the fenced perimeter at the existing Thompson Reservoir site near the Mountain Ridge subdivision. The two existing aboveground steel reservoirs at the site (2.5 mg and 4.0 mg) furnish water to the Thompson Pressure Zone. This pressure zone is located south of Highway 89. Pump Station No. 1 draws water from the reservoir's 18-inch pipeline and pumps to the Smith-Jones Pressure Zone, which surrounds the reservoir site. The pump station was constructed to provide adequate system water pressure and flow for this service area.

> Pump Station No. I consists of a concrete block building that houses three vertical turbine pump assemblies with room for a future fourth pump, a 16-inch bypass line, heating and ventilating equipment, and related piping, electrical, and instrumentation equipment. The building is automatically heated, cooled, and ventilated. In the event of a power failure, the bypass line will reduce surge conditions in the Smith-Jones Pressure Zone until the Hamilton Pressure Zone starts supplying water to the zone. The electrical system for the pump station is also sized for the ultimate addition of a fourth vertical turbine pump and its variable frequency drive unit.

> A separate room, accessible from the exterior only, was provided to facilitate the future addition of chemical injection equipment. Initially, however, there are no chemical injection facilities at the pump station.
>
> Pump Station No. 1 is one of three pumping stations used to supply water to various pressure zones. Figure 12.1 provides pressure variation information in these pressure zones with respect to the various pumps stations and their...

Specific equipment items are described in detail. The manual user is provided with all description information necessary to gain an understanding of the correct operation and maintenance of the equipment. The user may still be directed to manufacturers' literature for additional detailed information regarding specific equipment items.

Example:

> The three vertical turbine pumps located at Pump Station No. 1 are powered by 40 horsepower variable speed drive motors. Variable speed motors allow the pumps to be operated over a wide range of speeds. This allows a great degree of operational flexibility regarding pump selection and performance. Adjustable speed drive (ASD) units are provided to power each vertical turbine pump motor. These ASD units can be operated manually by the operator or automatically by a programmable logic controller (PLC). The PLC can automatically analyze each pump's performance and adjust its speed for the most efficient operation. In addition, each pump can be operated independently of the others, allowing for very flexible operating parameters.
>
> The master control panel contains the ASD units and the PLC. The PLC can be re-programmed to operate under a myriad of pumping conditions. The following are detailed instructions for re-programming the PLC and its...

Process or equipment operating procedures are described in detail. Relationships between the equipment or process in question and other related equipment/processes can also be discussed. Virtually all operations and maintenance personnel should gain a complete understanding of the operation of the equipment or process after reading the section. Theoretical operating procedures and abnormal and/or emergency operating procedures can also be covered at this level of detail.

Example:

> Each pump is equipped with a three-position hand select switch labeled ON-OFF-AUTO, located on the front of each control panel. When this switch is placed in the OFF position, the pump will not operate. When this switch is placed in the ON position, the pump will start in manual mode and its speed can be controlled by the manual speed control adjustment knob located on the front of each ASD panel.
>
> When the ON-OFF-AUTO switch is placed in the AUTO position, the pump is controlled automatically by the PLC located within the main control panel. The PLC will turn pumps ON and OFF and control their speed based upon system demand to meet the desired system PRESSURE SET POINT. When functioning in the automatic mode, the PLC is referred to as the "pressure controller." The pressure control system is located within the main control panel. The control panel has a three-position hand selector switch labeled MASTER CONTROL LOCAL OFF-REMOTE, a system pressure set point adjusting knob labeled PRESSURE SET POINT, a 0-100 psi pressure meter labeled DISCHARGE PRESSURE, a 0-20 psi suction pressure meter labeled SUCTION PRESSURE, a 0-6,000 gpm flow meter labeled FLOW, a reset push-button switch labeled ALARM RESET, and an alarm annunciator panel labeled ANNUNCIATOR.

Troubleshooting sections are common at a high level of manual detail. However, manual users are still directed to manufacturers' troubleshooting literature to obtain detailed information.

Example:

> If any of the vertical turbine pumps do not appear to be operating correctly or efficiently, the following list of troubleshooting procedures may be employed to aid in locating the problem:
>
> If the pump is making abnormal noises during operation, try to determine where the noise is originating. Inadequately lubricated bearings or seals can often cause excessive noise. Check that adequate seal lubrication water volume and pressure are being supplied to the pump shaft seal. Check that the motor has been lubricated at the proper intervals.
>
> If the pump's flow rate or discharge pressure is below normal, the pump suction valve may be partially closed or the suction piping may be partially obstructed. Check that the suction valve is adequately open.
>
> The equipment manufacturers' catalogs also contain troubleshooting information regarding most problems normally associated with their equipment. If these catalogs do not address a particular problem, call the equipment supplier or manufacturer and request this information directly.

Equipment lists generally provide adequate information at this level of detail. Manual users are directed to the manufacturer's literature regarding detailed specific equipment item and performance data.

Example equipment list:

Vertical Turbine Booster Pumps

Quantity	3
Pump Column Size	8-inch
Discharge Pressure Rating	76 psi
Available NPSH	58-65 ft
Required NPSH	13 ft

Number of Pump Stages	3
Design Capacity	1,500 gpm @ 118 ft TDH
Manufacturer	Super Pumps, Inc. 2000 North St. Railroad Anywhere, MA 78787 (555)123-4567
Supplier	Pumps, Inc. P.O. Box 0001 Pumps, MA 92222 (555)765-4321
Motor Type	TEFC
Manufacturer	Electric Motors, Inc. 123 Super Road Motors, CT 06001 (555) 232-2323
Supplier	Motors, Inc. 999 Rough Road Sacramento, CA 95005 (555) 333-3333
Horsepower	75
Type	Variable Speed
Speed (max)	1,770 rpm
Power	480v, 3 phase, 60 Hz
Frame No.	404VP
Motor Efficiency @ Rated Capacity	93%

While it can be difficult to determine the target audience and level of detail for a particular O&M manual project, following the steps outlined in this chapter will greatly increase your chances of success. Every O&M manual is unique, but each has a common thread as well. Remember that the overall goal is to convey the information as concisely and accurately as possible.

Chapter 2

O&M Manual Basic Components

Every O&M manual has at least a few major sections and subsections. This chapter explains what these sections and subsections are, and what information each should include. With input from the end-users, you can quickly compile the manual outline, style guide, and basic section layout for the manual. This chapter will cover the most common O&M manual components:

- Overall organization
- Outlines
- Table of contents
- Manual sections

Most easy-to-use O&M manuals have one thing in common—a basic, simple structure. If the sections are sequentially arranged and the text carefully formatted, the manual will be easy to read and understand. If the manual is put together haphazardly, information will be hard to locate and the manual will not be used effectively, or it will be shelved and not used at all.

Locating vital information within a technical document without an index or properly tabbed and annotated sections can be frustrating. The same can be true for O&M manuals. During a major process upset or equipment malfunction, the operator or technician needs to quickly find vital information that can help remedy the problem. If you have done your job, the reader will find the necessary information quickly and efficiently.

Overall Organization

The overall organization, or layout, of the manual is very important. Your end-user should be able to easily locate and understand information in the manual. When at all possible, put chapters and sections together in a seamless, logical fashion. For instance, most O&M manuals begin with a chapter covering a general description of the facility or equipment. Present the material with a gradually increasing level of detail. Beginning the manual with detailed explanations of process control and operations may confuse the end-user and intimidate staff trying to grasp basic plant information. If a manual is laid out in a logical, simple structure, the user will find it much easier to understand. As a result, the manual will be viewed as an "effective document."

Determining the most effective organization for a specific facility's manual can be difficult. Applying standard modes of operation to similar facilities does not necessarily mean you can use the same set of rules for organizing a manual for each. Manuals written for small facilities are usually limited in size and, therefore, are easier to organize. However, manuals written for larger facilities are often much more complex. Although the facilities may be similar in operations procedures, each may require a manual that is arranged differently.

O&M manuals can be arranged by:

- Facility processes
- Facility areas
- Major buildings or structures
- Equipment
- Service areas

Outlines

One way to arrange a workable organizational structure is to divide each chapter into several sections and subsections. In general, most writers don't use enough headings to break up passages of text. Not only do headers give the reader's eyes a break, they make finding information much easier. The end-user can skim headings and subheadings to locate specific topics. As a side benefit to you, the writer, they make revising the manual much easier.

You also need to insert section headings whenever the subject matter changes, even if it changes infrequently. Don't provide pages and pages of uninterrupted text.

Consult with the end-users that have the final say before determining the final organization of the manual. Several outlines should be discussed prior to compiling any sections of the manual.

Table of Contents

A critical component of the O&M manual is the table of contents, providing an exact listing of chapters and subsections. The users' first impression of the manual will most likely be formulated after looking over this section. Users usually do not read a manual from cover to cover, but use it as a reference. This makes an effective table of contents all the more valuable. No matter how organized the manual, a poorly structured table of contents can make the manual ineffective.

The table of contents can be color-coded with tabs for each section or subsection. Color coding allows the user to quickly flip to a pertinent manual section.

The table of contents enables the user to locate information quickly and easily. In addition, if the manual does not have an index, the table of contents becomes the sole guide for the user.

Manual Sections

Every facility or piece of equipment is unique. The following sections are a guide for both content and format for a typical facility O&M manual. Not every manual will contain all of these example sections; some contain more sections, some less, but most incorporate variations of those shown.

Equipment or Process Overview

Every O&M manual includes an overview section for each major process or equipment item. Generally, the section should provide the following information:

- Brief physical description of equipment, facility, or process
- Brief description of why the equipment, facility, or process is needed
- Brief description of what performance requirements are expected of the equipment, facility, or process
- Brief discussion of when the equipment, facility, or process was designed and constructed (optional)

Remember that the overview is *only a synopsis* and detail is minimal. Following is an example of a poorly written process overview:

WRONG

> The Winding River Wastewater Treatment Facility is an activated sludge secondary treatment plant comprised of primary sludge settling, secondary filtration, aerated sludge basins, and supplemented with an effluent filtration and disinfection process. It treats the wastewater from the community of nearby Smithtown, which is populated by approximately 5,000 persons.
>
> In an effort to meet strict state and federal pollution guide-lines, the plant removes organic and suspended material from the waste stream to the level provided by its various processes. The treatment processes substantially reduce the amount of pollution entering the Winding River.
>
> Before the treatment plant was constructed, large amounts of untreated wastewater polluted the salmon-rich waters. Once the wastewater is treated adequately, it is disinfected and dechlorinated, then discharged into the Winding River. Wastewater flows entering the plant do not vary widely during the year. After several efforts in the late 1970's and early 1980's to raise money for a new plant, the existing plant was finally designed and constructed in 1985.

Wastewater characteristics and design criteria are as follows:

Flow (mgd)

Average Day	2.5
Maximum Day	6.8
Maximum Month	3.8

BOD (lb/day)

Average Day	4,800
Maximum Day	8,600
Maximum Month	6,200

TSS (lb/Day)

Average Day	4,500
Maximum Day	8,100
Maximum Month	5,900

A table of equipment design criteria and a glossary of wastewater terminology are available in the appendices to this manual.

The previous overview demonstrates several errors. First, the opening sentence mentions every process at the plant. Inexperienced operations or maintenance staff may be turned off at such an overstated first sentence.

Second, manual users at this plant will probably know that the plant serves the neighboring community of Smithtown, that the population is about 5,000, and that the main reason for the plant is to reduce water pollution. Most, if not all, of the second and third sentences can be deleted.

Third, phrases like "large amounts of untreated wastewater polluted the salmon-rich waters," "treated adequately," and "flows entering the plant do not vary widely" are very subjective and can confuse the user. Such statements can even be taken out of context and be very misleading. The statements bring several questions to mind:

- How large were the amounts of untreated wastewater that polluted the river?

- What level of water treatment can be considered to be adequate?

- Just how much does the incoming plant flow vary?

If this information is to be presented, do so in a later, more detailed chapter or section.

Finally, the overview is not the place for design criteria or performance tables. Including such detailed information suddenly turns your operations manual summary into a detailed plant performance chapter. Following is an example of a more effectively written procedure overview.

BETTER

> The Winding River Wastewater Treatment Facility is a conventional activated sludge secondary treatment plant with effluent filtration. It treats wastewater primarily from the nearby community of Smithtown. The plant removes organic and suspended material from the waste stream to the level provided by secondary treatment followed by filtration. The wastewater is disinfected and dechlorinated before it is discharged into Winding River. The plant was originally designed and constructed in 1985. A table of equipment design criteria and a glossary of wastewater terminology are available in the appendices to this manual.

Although this section is shorter, it covers the same information without unnecessary, confusing detail. This same "short and simple" overview technique should be applied to equipment overviews as well. Again, an overview section should precede all major process sections within the manual. If the facility includes a number of treatment stages or processes, you must include a process diagram.

Equipment or Process Design Information

The equipment and/or process information an engineer uses to make design decisions is called *design criteria* or *design data*. This information can be very important when attempting to remedy process upsets or when upgrading or improving the facility. However, operators and/or maintenance technicians will probably not refer to the information frequently. Some design information lists may be more detailed than others. If the table of design information is kept concise, you should include a reference to the documents containing a complete list of information, such as engineering reports.

When the O&M manual is initially compiled, you can insert several blank pages after the table of design criteria and label them "Design Alterations or Modifications." The section reminds end-users to record design alterations and modifications in the document and gives them an area within the document to easily record the information. This type of simple record keeping can save an enormous amount of time and money when an engineering firm designs modifications to the facility.

Equipment or Process Control

The equipment or process control sections are usually the heart of any O&M manual, and you must ensure that these sections are both complete and accurate. The following information should be included:

- Introduction and summary of the equipment or process
- Theoretical and empirical information about the equipment or process
- Controllable and uncontrollable process variables
- Equipment configuration and characteristics
- Process overall characteristics
- Process calculations
- Process target values
- Process or equipment troubleshooting.

Each equipment or process control section should be preceded with a brief introduction. The writer can provide as much detail as necessary to accurately describe equipment or process control. Following is an example process control introduction for a wastewater treatment process:

> The goal behind any effective wastewater treatment activated sludge system is to develop and maintain a well-balanced sludge that will allow stable operation while

tolerating varying hydraulic and organic loads. The aeration basins provide the basic environment in which a
population of microorganisms can grow and be success-
fully maintained. The important elements that affect the
sludge growth and corresponding removal efficiencies include pH, alkalinity, temperature, dissolved oxygen, am-
monia, and biochemical oxygen demand (BOD) loading.

It is also important to control toxic substances that might
enter the activated sludge environment. These substances
(pesticides, chlorine, and heavy metals like copper, zinc,
chrome, and nickel) will affect the microorganisms by their
presence or by sudden changes in their concentration.
Therefore, close monitoring and control of all of these
elements are necessary to ensure a stable, efficient biological process.

This example provides the user with a substantial summary of the
process while limiting the amount of detailed process information.
Process control details are better placed in a more specific subsection.

Theoretical and empirical performance information generally apply
more to processes than to individual equipment items. Every process
(treatment, power generation, refining, etc.) has its basic foundation
in theoretical and/or empirical data. Information gained through the
application of various process control techniques can be quantified
and refined to make the process more efficient.

For instance, the process of refining crude oil can be described in
scientific and mathematical terms. The process can also be described
using empirical, or verifiable, methods. This section should include
both types of process information. Based on how the process should
theoretically perform, technicians can use actual performance data to
predict how the process will perform in the future. Past performance
information can also be used to prevent process upsets and other
problems.

Field verifiable data can be very useful for present and future plant
operations and design efforts. Include several blank pages at the end
of the section marked for notes to allow the user to record additional
information.

Every process system contains operational factors, or variables, that change continuously. Some variables, such as temperature or humidity, may vary only slightly. Others, such as flow rate or chemical reaction temperature, may fluctuate considerably throughout the process. The facility or equipment operator may be able to control many of these factors with equipment or process adjustments. For instance, adjusting water pump speed or pressure control valve settings can control pressure in a water service main. Such factors are called *control variables*.

Control variables should be categorized and described in this section of the manual. The number of control variables discussed in the manual often depends on the sensitivity of the process. For example, a process that dries newly formed concrete bricks would probably not be sensitive to minor temperature fluctuations. However, the disinfection exposure time for drinking water could be very critical.

You must list each control variable along with its effects on the process. Also include a description of how each control variable can be adjusted. Again, the detail of the process control variable section depends on the critical nature of the process. Process or operation factors that cannot be controlled or adjusted are sometimes called *uncontrolled variables*. Examples of these are wind direction, ambient temperature, sunlight, rainfall, humidity, etc. A description of uncontrolled variables pertaining to each process or equipment item should be listed along with a description of its possible effects.

Process Overview

The equipment or process overview section of an operations manual would not be complete without information about equipment description, configuration, and characteristics. Most facility processes contain a myriad of equipment types ranging from small items such as valves, to very large items such as pumps, reactors, and boilers. Large or small, each equipment item is a "link" in the process "chain." Any weak link can cause process or equipment upset or even failure. It is therefore very important for operating staff to become familiar with each equipment item and its role in the process.

You must include each definable equipment item, no matter how small, in this section. The information listed includes, but is not limited to, size, function, capacity, and configuration. Equipment items can be listed by type, such as valves, or individually, if only a few equipment items are discussed. For example, separate tables can be compiled for a listing of all valve numbers, location, function, type, etc. The manual user can then locate specific valves quickly and easily. The same technique can be used any time there is a large number of similar equipment items. Large equipment items are better described separately.

Most facility processes involve producing or treating something. For example, refining, treating, filtering, disinfecting, sterilizing, cooking, etc. The product undergoing treatment or change has many characteristics in itself. For instance, all crude oils possess many common characteristics before refining begins. Wastewater also has many unique characteristics. These attributes, though obvious, are nevertheless part of the process and should be understood fully by the user.

Product characteristics and their effects on the process should be listed and described thoroughly. Following is a description of water drawn from deep wells before treatment and disinfection:

> Water pumped from deep wells often contains large amounts of dissolved solids. The dissolved solids can be in the form of minerals such as sodium, iron, manganese, sulfur, etc. The water is often more alkaline than surface water, producing pH values of between 7.0 and 8.5. Well water can also appear to have various colors due to impurities.

Process Calculations

Facility operators must also have the skills to control equipment or processes. Computers can be used to automatically adjust equipment and therefore alter the process. But more often than not, the operator, using simple or complex mathematical equations, formulates process changes. These equations are called *process calculations*.

Process calculations are a very necessary part of operations and should be easily accessible and understandable by the manual user.

Some calculations will be very simple, such as the volume of a water-holding tank. Some, however, can be very complex. In this section you should present each major calculation separately, along with a detailed description of the calculation. If you feel you do not have adequate knowledge of the subject, consult an engineer or staff member who is familiar with the process. The facility operations staff should review the section carefully for accuracy of process calculations. Incorrect process calculations can cause simple processes to behave quite unexpectedly. Therefore, you need to review and verify process equations with a qualified and authorized facility staff person.

The following is a list of general rules for formatting equation information in the text:

- Equations should always be preceded by several blank lines and centered on the page.
- Equations should never be embedded within sentences.
- Complete definitions of all variables used within the equation should be included afterward.
- All information should be included on the same page, if possible.

Following is an example equation:

$$V = \pi r^2 h \times 7.48$$

where,
V = volume of tank in gallons
r = radius of tank in feet
h = height of tank in feet
7.48 = gallons of water in a cubic foot

This equation is used to calculate the volume of water in a cylindrical tank. For this calculation to be accurate, the tank must be truly cylindrical in shape. A sloped or depressed floor will cause the calculated volume to vary.

As you can see from the example, the equation is clearly set apart on the page from the main text. Also, commonly known and understood mathematical symbols, such as the Greek letter π, are normally not defined. However, if the letter, symbol, or number might be misunderstood or require explanation, include it in the definition list.

A paragraph is also included that describes the purpose and general use of the equation. The paragraph also points out variables that may cause errors or inaccuracies in the calculation. Provide this information to the user whenever possible. Following is an example of a more complex equation. The methods for presenting it on the printed page, however, are the same.

Example:

> The solids production rate (Y) of an activated sludge system is usually based on the pounds of total suspended solids produced per pound of total BOD removed. This parameter must be based on plant performance over an extended period of time in order to have any real validity. Most activated sludge systems produce from 0.4 to 1.0 pounds TSS for each pound of total BOD removed.
>
> Solids production rates generally vary with the sludge retention time (SRT). Plants operating at low SRTs (higher F/M ratios) usually produce more solids per pound of BOD removed. Lower solids production rates are found in systems operating at high SRTs (lower F/M ratios).

$$Y = \frac{(SS \times Q) + (TSS_w \times VOL_w) + ((MLSS_t - MLSS_y) \times VOL \times N)}{(TBOD_i - SBOD_e) \times Q}$$

where,

SS = secondary clarifier effluent total suspended solids (mg/L)
Q = daily average flow rate (MGD)
TSS_w = WAS total suspended solids (mg/L)
VOL_w = total WAS volume (mgd) (for all basins in operation)
$MLSS_t$ = today's MLSS (mg/L)
$MLSS_y$ = yesterday's MLSS (mg/L)
VOL = reactor volume (MG)
$TBOD_i$ = reactor influent total BOD, (mg/L)
$SBOD_e$ = clarifier effluent soluble BOD, (mg/L)
N = number of reactors in operation

The following is an example of a solids production rate calculation:

Given,

SS = 10 Mg/L
Q = 2.5 MGD,
TSSw = 6,000 mg/L
VOL_w = 24,640 gpd 0.024 MGD
$MLSS_t$ = 2,500 mg/L
$MLSS_y$ = 2,475 mg/L
VOL = 0.69 MG
$TBOD_j$ = 120 mg/L
$SBOD_e$ = 11 mg/L
N = 2

Calculate:

$$Y = \frac{(10 \times 2.5) + (6{,}000 \times 0.024) + ((2{,}500 - 2{,}475) \times 0.69 \times 2)}{(120 - 11) \times 2.5}$$

Y = 0.75 lb TSS produced/lb BOD removed

The preceding example includes a paragraph before the equation that explains the purpose of the calculation. After the equation and its variables are defined, an example is provided. Examples are a very important part of presenting such calculations, because they enable the user to "walk through" the calculation, thus taking the fear out of the process.

Process Target Values

The equipment or process operators must have a general idea of what process values to calculate—in other words, what values they're "shooting for." In this section of the manual, you should provide such information in the form of *process target values*. These target values give the manual user an idea of what calculated process values are.

Process target values are similar to process equations and should be treated with caution. As with equations, always defer questions about target values to the equipment or process control analyst/engineer for the facility in question.

Troubleshooting

Finally, all equipment or operating procedures must be evaluated from time to time by means of troubleshooting. Troubleshooting guidelines can be very helpful if the manual user is faced with solving an equipment or process problem. Although troubleshooting guides can come in many different forms, the guide must be easy to follow to be effective. Some troubleshooting guides are presented in the form of decision charts.

A common form of troubleshooting guide is the problem-cause-solution table. This traditional approach to troubleshooting is effective and easy to understand. It provides the manual user with a list of problems that might occur, along with the possible cause and remedy. Possible causes are usually listed in decreasing likelihood. Following is an example of a simple problem-cause-solution troubleshooting table for a centrifugal pump.

Problem	Possible Cause	Solution
Centrifugal pump is not pumping at full capacity	Suction or discharge valves are not fully open	Open suction and discharge valves to full open position
	Air in pump volute casing	Bleed air from volute by opening top air release valve while pump is running
	Suction or discharge lines partially restricted	Inspect suction and discharge lines for obstructions

Routine Checks

Routine checks of equipment and processes are a necessary part of most facility operations. Staff must check equipment periodically for abnormal wear or signs of damage. Also, they must sample and review processes for efficiency and proper function. Routine checks can include the following:

- Equipment checks for malfunction, improper adjustment, abnormal wear
- Periodic adjustment of process or equipment control variables
- Equipment preventive maintenance checks and adjustments
- Process sampling

In this section of the O&M manual, include routine check guideline information such as a description of the check, the primary purpose for the check, and how often the check should be performed. The facility owner often develops this type of information with help from the design engineer. However, laws and/or government regulations dictate many routine check procedures. Some regulations even require that the check procedures be documented in a particular format. Therefore, be sure to include all the required routine check information in the section. After the section is compiled, have a qualified staff person review it for accuracy and completeness.

Group the routine check information into similar tables or lists. For example, list all sampling information together in one table and preventive maintenance checks in another. Following is an example of such a list:

Sample Description	_Frequency_
Influent Water	Every 6 hours
Primary Clarifier	Every 4 hours
Primary Sludge	Every 6 hours
Secondary Clarifier	Every 4 hours
Effluent	Every 6 hours
Receiving Water	Every 6 hours

Component Lists

In a separate section of the manual, list all major components within a process. Information such as equipment name, manufacturer, model number, supplier, nameplate data, etc., can be very useful to the facility operator or maintenance technician. This information should be easy to locate and understand. Following is an example:

Pump Information

Process Designation	P-4-1-1
Type	Vertical turbine
Manufacturer	Excellent Pump
Model No.	456-02
Impeller type & diameter	Mixed flow; 18-inch
Pump speed	1,200 rpm
Pump output	5,000 gpm @ 26 ft TDH
Motor horsepower	60

This information about the equipment can be presented in detail or as generally as necessary. If you are not sure about the accuracy of equipment information, consult the manufacturer about current information along with any new maintenance and operation updates.

Many maintenance management software programs allow you to store this equipment information on your computer. Use the maintenance management program as the primary source to store equipment information. You can refer the manual user to the maintenance management program for needed information.

Instrumentation and Control Systems

Today, even the most modestly designed facilities are equipped with advanced control instrumentation systems. Larger, more complex facilities have sophisticated computer monitoring and process control capabilities. Generally, the overall usefulness of this technology has increased in the last few decades while the cost has generally decreased. Instruments are now much easier to maintain, are more dependable, and are often self-diagnosing.

So, why do we need operations manuals for these systems? The answer is simple: Most plant operations and maintenance staff have had difficulty keeping up with this growing technological trend. While instrumentation manufacturers and suppliers are often willing to help with installation and troubleshooting problems, day-to-day operations and maintenance of the equipment is challenging.

If you plan to include a section about instrumentation and control systems in your manual, make sure it summarizes all the manufacturers' literature. Some computer control systems require several large reference operations manuals for a single system. It would be nearly impossible, and entirely unnecessary, to transfer all manufacturers' information about such a system into the O&M manual.

Your goal, instead, is to provide a basic understanding of the system and its intended operation. Every type of system, be it mechanical, electronic, etc., must have a basic, logical structure. To be able to write an effective summary or introduction in the manual, you must have a basic understanding of the system. If this is not possible or practical, consult the designing engineer and/or the manufacturer of the instrumentation system for assistance.

Most instrumentation and control systems are designed with primary control logic configurations referred to as *control loops*. Control loops describe stand-alone parts of the instrumentation and control system. Describe their relationship to other loops and various other parts of the system separately. This helps prevent the user from becoming confused about the role a particular control loop or subsystem plays in the process.

The users more easily understand information about systems if it is presented one item at a time. For example, discuss the control panels for a particular process in one section and include the following:

- Process and instrumentation control number
- Item description
- Process loop number
- Purpose and general operation description
- Troubleshooting guide
- Manufacturer's information location
- Routine maintenance
- Calibration guide

Remember to keep these sections very general and refer the user to the manufacturers' literature for more technical information.

Alarm Lists

Alarm lists are also useful for operators and maintenance staff. Most instrumentation and control systems have alarms for various conditions within the process. These alarms can affect worker safety or process integrity. Discuss each alarm point separately and thoroughly. You can refer the user to the manufacturers' literature concerning troubleshooting, maintenance, and calibration of alarm points. However, provide a complete description of each alarm, possible causes, and reconciliation.

Process Control Drawings

Process control drawings are often included in this section of the manual. These drawings can be very informative to operators and technicians because they describe the overall process using diagrams. Use industry standard process and instrumentation (ISA) symbols whenever possible. Anyone viewing the drawings can then easily read and understand them. (The drawings should be placed within, or at the end of, the section in which they are discussed.) Be sure to reference all drawings within the text.

Manufacturer Information

Many manuals contain only references to manufacturers' literature. When appropriate, though, include the literature in the manual in its original form. Avoid including only part of a manufacturer's manual or brochure and always advocate manufacturers' methods and recommendations. Most manufacturers don't mind if their equipment literature is copied with permission, and requesting permission to reproduce updated information often yields new information about the product. Most manufacturers and vendors are more than happy to provide you with information about their products, and often produce product information already pre-punched for three-ring binders. Many vendors even provide information on their web sites.

Separate manufacturer information from the rest of the manual by using divider tabs. These tabs can be marked to match the table of contents so the user can find the information easily. The sections can also be color coded to aid in information location.

Manufacturers' literature often provides information about more than one equipment model number within a product line. For instance, a pump manufacturer might provide information on several of its pumps in one catalog or manual. To avoid confusing the user, cross out non-applicable information with a marker, taking care not to make any marked out information unreadable. In this way, information that applies to specific equipment items can be quickly located and information about other models can still be readable.

Some facility construction contracts require that the contractor and/or equipment suppliers provide a complete list of all manufacturers' literature for the project. You can group the information by marking out non-applicable information and using tabs and labeled, three-ring binders.

Startup and Shutdown Procedures

Every piece of equipment or process must have clear startup and shutdown procedures. Without such procedures, the technicians would question the correct operation and performance expectations of the equipment. Therefore, staff must have access to proper methods of safely starting and stopping the equipment. The listed procedures can also serve as training tools.

Write the startup and shutdown procedure sections in easy-to-follow, easy-to-understand steps. Your instructions on how to properly and safely operate equipment must be very detailed. Following is an example of a poorly written startup manual section, followed by a revised version:

POOR

> Startup Procedures for Water Distribution Pump P-4-1-1:
> First, check that the pump's suction and discharge valves are open. Next, turn pump on. Adjust pump variable speed control for desired speed. Listen for strange noises coming from the pump. Next, check that the pump discharge pressure is normal.

Presented with the above example, most manual readers will have immediate questions about the startup procedure:

- Should the suction and discharge valves be opened 100 percent, or only partially?
- How do you turn the pump on?
- Where do you turn the pump on?
- How do you adjust the pump's speed?
- Why should I listen for strange noises?
- What noises should be considered strange?
- What do I do if I hear strange noises?
- What is normal discharge pressure for this pump?

As you can see, this startup procedure example creates far more questions than it answers. Most operators prefer a step-by-step, or standard operating procedure (SOP) approach to procedures. SOPs are often more clear, easier to prepare, and easier to follow. Consider the following revised version:

BETTER

Startup Procedures for Distribution Pump P-4- 1 -1
The following steps should be performed sequentially:

1. Verify that the 12-inch pump suction butterfly valve SV-4-1 -1 is full OPEN.
2. Verify that the 8-inch pump discharge butterfly valve DV-4-1 -1 is full OPEN.
3. Verify that pump seal water supply is available and a minimum of 30 psi.
4. Place pump variable speed control to 0 percent speed. (Pump variable speed control is located on AFD panel AFD-4-1-1 near the pump).
5. Turn pump ON-OFF control to the ON position. (Pump ON-OFF control is located on control panel LCP-4-1-1 near the pump).
6. Adjust pump speed control (located at panel AFD-4-1 - 1) to 100 percent speed.
7. Observe pump discharge pressure. Discharge pressure should rise to between 50 and 100 psi within 30 seconds. If pump discharge pressure falls above or below this range, turn pump OFF immediately and refer to section TROUBLESHOOTING.
8. Listen for abnormal noises from pump. Pump should be turned OFF if loud grinding or knocking noises are heard. See section TROUBLESHOOTING.

The revised example gives a clear set of steps to follow for starting the pump. Words like "ON" and "OFF" have been capitalized to emphasize their importance in the step. Control panel locations and functions are provided, and expected pump discharge pressures are listed. Applicable references are also made to the troubleshooting section for resolving problems.

Make this section clear by including the item designation number after references to equipment items such as control panels, equipment items, and valves. Startup procedures for similar equipment items can be grouped together, but any differences in physical description or operational procedure should be clearly stated.

Describe startup procedures for entire processes in much the same way. The procedures will probably, however, be described in much more detail than shown here. The key is to provide a clear and concise description of startup and shutdown steps that anyone responsible for the equipment or process can easily follow and understand.

Manufacturers' startup and shutdown procedures are often limited to the equipment item in question and not to systems or subsystems. If the functional relationship between the equipment item and its related systems or subsystems needs more clarification, consult with the facility design engineer.

Most equipment items have normal modes of operation; however, some items can be operated in "out of the ordinary" modes. For example, suppose an oil refinery's crude oil transfer pump is normally operated to transfer oil from one tank to another. Under abnormal conditions, pump flow can be redirected to provide auxiliary pumping to other tanks. Procedures such as these need to be documented just as though they were normal operating procedures. Technicians often develop such procedures after years of operating the equipment. Mark these sections clearly as *nonstandard procedures* and have operations management review them for accuracy.

Emergency Procedures

Emergency procedures, usually defined by management, are unique for each equipment item or process. However, it is a good idea to provide the manual users with an easy-to-locate section about emergencies. Describe in detail all specific procedures to be followed for unique emergencies.

Write this section clearly and tab with a brightly colored divider. Also, highlight the emergency section listing in the table of contents and index.

Clearly describe procedures during and after accidental spills involving hazardous materials. The importance of having staff at all levels informed about how to report procedures for spills cannot be overemphasized. Prompt reporting ensures minimal contamination, health hazards, and damage to the environment. Being negligent in reporting and properly responding to such spills can result in severe regulatory penalties, not to mention the potential damage to the facility's public relations efforts. In some case, individuals can be held responsible for proper handling and reporting of spills.

Include in this section a discussion of the federal, state, and/or governing agency laws and/or regulations related to hazardous spills. Clearly outline the responsibilities and liabilities of the facility owner and describe the penalties for violations. Outline reporting requirements and procedures for spills and list routine and emergency telephone numbers for immediate reporting. Provide sample reporting forms and instructions for completing them. *Clearly spell out* the responsibilities of facility staff for reporting spills.

Federal, state, and/or agency guidelines for reporting and responding to such emergencies are subject to constant revision. Therefore, include a note in this section to inform the user of possible revisions.

Preventive, Corrective, and Predictive Maintenance Procedures

Preventive maintenance procedures are those activities performed routinely to extend the life of, or increase the performance of, equipment items. These maintenance procedures are performed in hopes that the equipment will have a longer life with fewer break-downs and repairs or replacements. An example is the weekly inspection of an air compressor's drive belts, greasing of a motor bearing, or the monthly adjustment of a pump shaft seal.

Corrective maintenance procedures are activities performed because of partial or complete equipment failure. An example is the replacement of a broken drive belt or pump bearing.

Predictive maintenance procedures are those used to predict the performance of an equipment item. For example, by tracking a particular type of equipment item over a long period of time, a similar item's performance and life expectancy can be estimated.

All of these maintenance procedures are a very necessary part of any facility's maintenance program. However, the scope of this book prevents a detailed discussion of maintenance procedures. The general presentation and format of such information within the manual should be clear and concise.

Because each represents a different facet of maintenance, place preventive, corrective, and predictive maintenance discussion in separate sections:

- List preventive maintenance procedures by equipment item, procedure, and frequency.
- List corrective maintenance procedures by equipment item and procedure.
- List predictive maintenance procedures by equipment item.

Many maintenance management software programs allow input and monitoring of preventive, corrective, and predictive maintenance information. If such a program is in place, the O&M manual *should not* contain redundant data. Refer the manual user to the maintenance management program for more detailed maintenance information.

Safety

Carefully compile the manual sections regarding safety. Incorrect recommendations and advice can create a hazardous situation and unnecessary liability. Refer the manual reader to manufacturers' literature regarding specific equipment safety recommendations. The facility or equipment owner should always review and approve safety recommendations. Following is an *example* safety section with very generalized recommendations:

Example:

GENERAL SAFETY GUIDELINES

The following guidelines will reduce the likelihood of employees being injured on the job:

1.   Observe all written and oral safety rules and be aware of the particular hazards surrounding your job.

2.   Do not start a task until you have received and fully understand the instructions.

3.   Immediately correct or report to your supervisor any hazardous conditions, unsafe equipment, or unsafe working practice.

4.   Report all injuries or accidents immediately to your supervisor.

5.   Drink only from water fountains. Other water sources within the plant may be unsafe.

6.   Do not run. Watch for and avoid slippery or congested areas.

7.   Do not ride on or operate any moving equipment unless it is part of your job and you have been instructed in its proper operation.

8.   When operating moving equipment, observe all traffic signs, speed limits, and parking regulations.

9.   Do not wear loose clothing or carry rags in your pockets. Cloth may become caught in equipment and cause personal injury.

10.   Use protective equipment such as goggles, hard hats, gloves, and respirators whenever warranted or required by the tasks.

11.   Do not operate any equipment unless all safety guards and safety devices designed for that equipment are in place, except as permitted in written maintenance or emergency operation procedures.

12.   Shut off equipment before cleaning debris from moving parts.

13.   If it is necessary to remove safety devices, handrails, manhole covers, or related items, warn fellow employees.

14.   Keep all tools in good repair and ensure that you use tools appropriate to the work being performed.

15.   Do not pass under or work beneath fellow employees unless a task requires doing so. Never enter a wet well, tank, or basin until all precautions have been taken to ensure safety.

16.   Practice good housekeeping. Immediately clean up any grease, oil, or hydraulic fluid that may have spilled or leaked from the equipment. Do not use gasoline to clean up oil and grease. Keep all passageways, aisles, stairs, and exits clear of tools, equipment, and other materials.

17.   Do not consider a job finished until you have made conditions as safe as possible for the next person.

Glossary

While not a required part of operations manuals, a glossary provides a clearer understanding of the equipment or processes. A glossary should always be included, however, if the manual is to be used for training. Following is a short excerpt taken from a wastewater treatment plant manual:

Example:

GLOSSARY

Absorption: Taking in or reception of one substance into the body of another molecular or chemical action, and distributed throughout the absorber.

Activated Sludge: Sludge particles produced in raw or settled wastewater (primary effluent) by the growth of organisms in aeration tanks in the presence of dissolved oxygen. The term "activated" comes from the fact that the particles are teaming with bacteria, fungi, and protozoa.

Activated Sludge Process: A biological wastewater treatment process in which a mixture of wastewater and activated sludge is aerated and agitated. The activated sludge is subsequently separated from the treated wastewater (mixed liquor) by sedimentation and wasted or returned to the process as needed.

Adsorption: To gather (a gas, liquid, or dissolved substance) on the surface or interface zone of another substance.

Aeration Basin: The tank where raw or settled wastewater is mixed with return sludge and aerated.

Aerobic: A condition in which "free" or dissolved oxygen is present in the aquatic environment.

Aerobic Bacteria: Bacteria that live and reproduce only in an environment containing oxygen that is available for their respiration (breathing), such as atmospheric oxygen or oxygen dissolved in water. Oxygen combined chemically, such as in water molecules, H20, cannot be used for respiration by aerobic bacteria.

Index

An index is more than just a list of words at the end of the O&M manual. It is a reference tool for the user to quickly locate necessary information. The index can often "make or break" the usefulness of a manual.

Today's word processing programs can compile indexes very quickly. However, a computer can't distinguish the importance of a word with regard to the text. So, when preparing the index, choose words and terms that are likely to be referenced. One way is to let the computer build the list of words. Then delete those words or phrases you don't wish to include in the index.

Include equipment numbers in the index to make it easy to locate other similar equipment numbers. For instance, the user could find all occurrences of the pump designation number P-2-1 -1 by looking through the "P" section of the index. If the entire manual document is produced on a computer, the index can be easily updated.

Chapter 3

Estimating, Scheduling, and Producing the Manual

Producing an O&M manual can be a challenge. All activities must be coordinated properly and the manual production team must work together to get the job done on time and within budget. This chapter covers the estimating, scheduling and production of a typical O&M manual.

Before the actual writing and production of an O&M manual can begin, you need to estimate the time and cost required to produce the manual. Even though everyone would prefer a manual that covers every facet of a facility's operations, its scope and content are usually limited by the available budget and time.

Two typical questions asked by an experienced writer are: (1) How much time do I have to produce the manual? (2) How much will it cost? This approach is often incorrect. The primary goal should be to provide the best O&M manual possible, while meeting the scope of the project for the least cost.

Therefore, the questions that should be asked are:

- What kind of manual does the user want and need?
- What is the scope of the manual?
- How many drawings and other graphics items will be included in the manual?
- How many writers are available to work on the project?
- How many copies of the manual will be needed?

Without answers to these questions, you are likely to flounder through the project. Jot down every possible question about the project and find the answers *before* you begin the manual. You will be more informed and feel much more in control when the project flows smoothly.

So, what do you do when management asks you to prepare a manual with limited budget and resources? First, develop a reasonable time and cost estimate using the steps in this chapter. If your estimate matches the original time and budget, the project should go well. However, if your estimate exceeds the original time and budget constraints, you need to talk to the project manager. Go over the manual's scope to make sure you're not including unnecessary work. Then, try to stress the importance of an accurate, well-written document. If the original budget and time constraints cannot be changed, make the project manager aware of the "corners" that must be cut in order to meet these constraints. Often, you will have to consult the end-user to determine which areas of the manual must be sacrificed to meet project constraints.

Estimating Time

Every O&M manual project varies in text layout, complexity, graphics, length, etc. However, there are set guidelines you can use to estimate the time needed to complete a project. First, you could determine the approximate time required for your project by compiling past manual projects in an electronic spreadsheet. Second, you could multiply the average time per page for all previous projects by the proposed manual page count. Third, you could multiply the average man-hours per page for each task by the proposed page count. Consider the following example:

> The average time required per page of text for all previous manual projects is 1.8 man-hours. To find the approximate time required for a proposed manual, multiply 1.8 man-hours by the proposed manual page count (250). The total time required for the proposed manual is 450 man-hours.
>
> To estimate the time required for editing, multiply the previous project average man-hours per page for editing (.15) by the proposed manual page count (250).

The total time for editing is 37.5 man-hours.

Electronic spreadsheets such as Microsoft Excel and Lotus 1-2-3 can save hours of calculation time when doing "what-if" scenarios with various project data. For example, you can plug in varying man-hours for tasks such as "editing" and see what happens to the project's bottom-line totals. Keeping a database of past manual project information in the form of electronic spreadsheets can save considerable time and expense when preparing future cost estimates. Having readily available information can help when you need to justify your time requirements at project meetings.

If past project information is non-applicable, questionable, or nonexistent, you will need a more conventional estimating approach. First, make a list of all the tasks required to complete the project such as:

- Client meetings
- First draft
- Editing
- Field research
- Manual review
- Graphics production
- Printing

Then, roughly estimate the manual's page count. To be safe, add 10 percent to the estimated page count. Next, ask each person involved in the manual project how long it will take to complete his or her tasks. Make sure you question each person about other ongoing projects that might cause scheduling conflicts. Be sure to provide each with an adequate description of the scope of the proposed manual. If the time requirements sound unrealistic, ask each to break tasks down into subtasks along with time estimates for each.

Estimating Cost

After the time estimate is completed, you can begin estimating the cost for the manual project. Every O&M manual project has two cost categories:

- **Labor**—cost of people working on the project, such as writers, editors, graphic artists, etc.
- **Expenses**—cost of materials and services, such as paper, computers, printers, reproduction, etc.

Labor cost can be determined by multiplying the estimated man-hours per task by the individual's labor rate, then totaling the costs for each task. Remember to include *all overhead costs* into the labor rate. The labor rate should be the amount billed against your project per man-hour.

Expenses can be estimated by totaling all expenses reasonably expected to be incurred during the project. If all expenses are not known at the time the estimate is prepared, you can apply a contingency percentage of approximately 10 to 20 percent.

Scheduling the Project

Since many of the tasks may be accomplished concurrently, a time line or activity bar chart must be developed to accurately estimate the project schedule. Smaller projects may require simple, even "mental," bar charts, while larger projects may require project scheduling software.

When developing a schedule for producing the manual, determine the order in which tasks must be completed. Every project has certain tasks that must be completed before another task, or tasks, can begin. For example, editing cannot be completed until the writing task is done. Therefore, "writing" will occur before "editing" on the bar chart. Also determine which tasks can be accomplished simultaneously. After all tasks are arranged in a logical order and a reasonable time estimate is assigned to each, you can determine an overall project schedule.

After a project schedule is completed, the project team should review it. This review gives everyone involved an opportunity to see how their part of the project "fits in" with the overall manual production plan. Any changes to the schedule could easily be made at this time.

The Manual Production Team

After the cost estimates and schedule are completed, select a writing team. One individual often can oversee many smaller manual projects. This person writes, edits, reviews, and produces the manual virtually single-handedly. Larger manual efforts, however, can employ teams of writers, editors, and graphics experts. Most manual projects fall somewhere in the middle.

For example, assume you are assigned to manage a 300-page O&M manual project. You have at your disposal several engineers, editors, graphic artists, and printing assistants. Also assume that the budget and time allocated for the project is barely adequate.

First, as the project manager, you should arrange a brief meeting of all key persons involved in the project. This group becomes the Manual Production Team. Inform everyone of the requirements of the project and assign individual tasks. Also discuss project milestones and budget and time constraints.

Don't worry if some team members are pulled off in the middle of your project to work on more urgent matters. Often, these members can, and should, be replaced with other competent persons. Make every effort, however, to keep the project on schedule. As a colleague once told me, "Show me a project that's late, and I'll show you a project that's over budget!"

Developing the Outline

Before work can begin on a draft of the manual, you must develop a usable outline. The outline is your road map of the content and scope while writing the manual.

Developing good manual outlines doesn't have to be difficult. Most facility operations manual sections are divided into specific subject areas. Brainstorming is often used to create outlines. First, list as many topics as you can about the facility. For example, a manual for a water treatment plant might be structured as follows:

1. Introduction
2. Treatment Process Overview

3. Unit Processes
4. Monitoring Requirements
5. Staff
6. Safety

Initially, the order in which these sections are listed is not important. More important is that they have been included; a logical order will evolve. Once this hierarchical order is established, break each section down into its logical subparts. For example, the section on Unit Processes can be expanded as follows:

3. Unit Processes
    A.   Water Wells and Pumping
    B.   Lime-Softening System
    C.   Filtration
    D.   Disinfection
    E.   Chemical Feed Systems

These sections can be broken down into:

1. Unit Processes
    A.   Water Wells and Pumping
          1. Purpose
          2. Well Construction
          3. Operating Conditions
    B.   Lime-Softening System
          1. Process Description
          2. Operation

    C. Filtration
          1. Process Overview
          2. Purpose
          3. Process Description
          4. Filter Operation
          5. Instrumentation and Control
          6. Troubleshooting
          7. Electrical
    D. Disinfection
          1. Disinfection and Trihalomethane Control
          2. Process Control
          3. Operation
    E. Chemical Feed Systems
          1. Lime Feed System
                a. Process Description
                b. Operation

        c.   Pump Calibration
2.   Alum Feed System
      a.   Process Description
      b.   Operation
      c.   Pump Calibration
3.   Hexametaphosphate Feed System
      a.   Process Description
      b.   Operation
      c.   Pump Calibration
4.   Polymer Feed System
      a.   Process Description
      b.   Operation
      c.   Pump Calibration
5.   Sodium Hypochlorite Feed System
      a.   Process Description
      b.   Operation
      c.   Pump Calibration

This is not an outline for the entire manual, but only for the section entitled **Unit Processes**. A similar structure applies to the other major sections of the manual. Remember, the outline can change considerably as the team offers more input. If you're not familiar enough with a particular facet of facility operation, consult the owner or design engineer, who can offer valuable suggestions for manual content.

Monitoring Progress

Regardless of the manual size, its development and production must be monitored to keep the project on schedule and within budget. To determine of you're on schedule, check the original time and cost estimates prepared before the project began. For example, suppose your original plan was to complete the first draft of the manual in six weeks, but in reality the project has been underway for six weeks and the first draft is only two-thirds complete. You're behind schedule.

Don't wait for the project to "correct itself." It won't! You have to make a course correction. Contact the team member responsible for the first draft to find out when it can be realistically completed. Offer the team member assistance. Be aware that your budget will begin to escalate because more effort is now required for this task. You may need to trim a subsequent task back in order to meet budget.

Most software scheduling programs can also track progress and costs for a project. Some programs can produce daily, weekly, or monthly progress reports and graphs concerning cost and scheduling. These features can help project managers find and correct "trouble spots" early in the project.

Gathering Information

Very few writers know everything about a facility or equipment item. Yet, they are often expected to produce an accurate, detailed O&M manual on the subject. How does a technical writer perform this magic trick? The answer is that the writer must either become an expert on the subject or know where to find the experts. He or she must know enough about the equipment or facility to be able to ask questions and interview persons that design and operate similar facilities and/or equipment.

If you're asked to manage an O&M manual project regarding a facility you know very little about, you may do less writing and more managing. On the other hand, you might be the sole writer on a manual project in which you're familiar with the facility. In any case, you'll most likely be perceived, in the end, as the "knowledgeable source."

At times, you will need to discuss operational and maintenance details with a member of the design engineering team, the equipment manufacturers, or both. Don't be intimidated. Set up interviews with them only after you're sufficiently prepared with a list of pertinent questions. To accurately convey this information to the user, get as detailed answers to your questions as possible.

Remember, document *when, where,* and from *whom* the information is obtained. Documentation references the source of the information and provides a database of knowledgeable individuals you can call upon later. Instead of an interview, you can send a letter of inquiry or telephone or e-mail. Always include in your message the date by which you need the information.

If the accuracy of your work is ever questioned, check to make sure you've recorded the information correctly. Often you'll find that the information is more subjective than incorrect. Subjective information must be noted as such in the O&M manual.

Another technique for learning about the subject or facility is to review manuals, articles, or periodicals on the same subject. For instance, you might be able to construct a better outline for you manual by reviewing a manual assembled for a similar facility.

Style Guide

One of the primary responsibilities of the project manager is to provide the production team with a *style guide* early in the project. A style guide could be a one-page memo or bound volumes defining all aspects of manual production.

Many government agencies and municipalities have developed their own style guides for manual and report production. Many facilities, however, do not have such guidelines. It is, therefore, important to that you ask the client if a particular style guide is to be used for a particular manual project.

The manual production style guide needs to define the following elements:

- Brief description of the manual project
- Word processing format
- Computer file naming
- Page numbering, headers, footers, and graphics numbering
- Typesetting
- Text heading hierarchy
- Level of detail
- Manual structural format
- Style conventions
- Drafting and graphics standards

Description of the Project

Many of the individuals involved in the production of an O&M manual will be acquainted with the facility or the equipment. Some may be experts on design parameters, operational characteristics, maintenance requirements, safety concerns, etc. Some, however, may know very little about the system or facility being documented. The style guide is a way to introduce the facility or equipment being documented, the manual's purpose and scope, and the targeted user. The following is an example of a style guide preface:

> This style guide was developed to facilitate the production of an accurate, consistent, and effective operations and maintenance manual. The scope, organization, format, level of detail, and style guides are the consolidated work of the operations staff of the facility.

Word Processing Format

Most documents today are developed using word processing programs on computers. These programs are efficient, easy-to-use, and provide a means for storing documents for long periods of time. The ease of editing, reformatting, and printing documents makes word processing software a necessary tool when producing O&M manuals.

Word processing programs, like most other technologies today, are produced by many different manufacturers and come in many different versions. Most facilities have a preference as to what word processing software they prefer. However, if there is no preference, a common application should be used. This will make future editing and formatting by other individuals much easier.

The style guide should include a brief description of the word processing software to be used along with instructions about file handling. Following is an example of a guide for word processing format:

> The word processing software to be used during manual production is Microsoft Word Office 2000. The files should be stored in the G:\USERS\MANUAL directory on the dedicated file server USR4. When saved to this directory, the files will be backed up automatically every night at midnight. For information regarding the software or the file server, contact Computing Services.

Clarifying how files are to be stored is very important. If several team members will work on the project at the same time, individual word processing files could be located anywhere. Having a style guide enables the project manager to locate and review the progress of all project files. All word processing files should be saved to a recognizable directory on a local hard drive and backed up frequently.

Naming Computer Files

Word processing files that pertain to your O&M manual project should be named properly. The style guide should explain the file naming standards to be used. The standard will help to locate files later—even years later—if additional editing or updating is required. One very effective method of naming files is to "code" each section of the file name. For example, the first four characters of the file name might represent the project, and next four characters might represent the chapter or section of the manual. Finally, the file extension can indicate whether the file is a draft or final. The following illustrates this type of numbering system:

File Name: CHAM0001.DFT
"CHAM" represents Chamberlain Petroleum Plant O&M Manual
"0001" indicates Chapter 1
"DFT" indicates a draft version

Although the standard for naming a computer file is arbitrary, a consistent approach is imperative. Whatever standard is used, make sure it is logical and easily recognizable.

Make sure to print the computer file name on the document pages themselves for easy reference. The reference should include the file name, revision date, and initials of the author in a page footer.

Page Formatting

The style guide should provide the manual production team with requirements for fonts, margins, hyphenation, tables, page numbering, headers, footers, and figure/table numbering. The manual will then have continuity and a uniform look and feel. The following is an example of format requirements:

Margins:	Left	1.2 inches
	Right	1.3 inches
	Top	1.0 inches
	Bottom	0.7 inches

Tabs set every 0.5 inches
Left justification
Right ragged
Times New Roman 12 pt. font
No hyphenation

Page numbering: section-page (Example: 2-1)
Tables numbered consecutively section.number (Example: 4. 1)
Figures numbered consecutively section.number (Example: 2.3)

Creating Headings

Text headings and subheadings give the O&M manual pages a professional look and allow the user to easily locate information on the page. A style guide should give clear direction along with examples as to the type of headings to use. Certain typefaces are always used for all chapter titles, different typefaces for section subheadings, and so on. The key is to use headings consistently.

Determining Level of Detail

The manual's level of information, or *level of detail*, needs to be determined before writing begins. The style guide should clearly spell this out for the writing team. Chapters and sections written in varying levels of detail can confuse the user. Therefore, consistency is very important. Sample sections should be given to writers to give them an idea of what may be too little or too much detail. Following is an example of the approximate level of detail in a pump control panel location section of a manual:

> Pump Control Panel Location
> The chemical feed pumps are locally controlled at field control panels located in the Chemical Building. These panels are numbered FP-2-1 and FP-2-2 and are located more specifically on the walkway adjacent to the chemical storage tanks. The walkway can be accessed from the lower level of the Chemical Building.

Manual Structuring Format

The overall format of the manual needs to be described in the style guide. Each chapter or section should be formatted consistently with similar headings and subheadings. A set format gives writers a pattern to follow. A manual with a consistent format is much easier to read and understand. Following is an example:

> Process Description
> Equipment Location
> Motor Control Center Location
> Normal Operation
> Startup
> Shutdown
> Maintenance
> Safety

Following Style Conventions

The style guide should include a list of *style conventions*. Style conventions are rules the writer must follow in order to produce a consistent document. Examples are numbering tables, figures, etc.; numbering the steps in a procedure, (e.g. 1, 2, 3); and using all capital letters for terms like ON, OFF, AUTO, START, START, etc. These style conventions must be clearly listed and used in examples when possible.

Drafting and Graphics Standards

Graphics standards vary widely from one manual to the next. Many facilities already have graphics standards that they may want used in the manual. Whatever the case, you need to be consistent so as not to confuse the user when using note call-outs, numbering figures, using borders, etc. The manual will look professional and the information will be easier to understand.

Chapter 4

Desktop Publishing an O&M Manual

With word processing programs and high quality laser printers, it's easier than ever to create professional looking documents. The term *desktop publishing* refers to the production of documents using computer software programs designed to provide results comparable to professional printing services. Scanned photographs and other graphics can be added easily.

While not every O&M manual should be produced using desktop publishing methods, doing so can make the document easier to read and understand. The use of icons and symbols in the manual can alert the reader to important information. This chapter covers tips and suggestions on how to enhance the O&M manual.

Manual Format

Manual format can be defined as the "look and feel" of the document. Factors such as readability, references, page layout, page numbering, margins, type sizes, and graphics are all part of the manual's format. Cover art also plays a role in the overall format of a manual.

The manual's format is often governed by the funds available to produce it. Plain typefaces and formats are generally less expensive to produce. However, the simpler versions are less likely to hold the interest of the reader.

It can also be expensive to change formats in the middle of manual production, especially when major changes are required. Therefore, the manual's format and general production direction should be decided upon early and adhered to throughout the project.

Readability

What is readability? The dictionary defines a readable document as one that is "easy or interesting to read." While not all operations manuals may be interesting, they should be easy to read. People are often selective when it comes to choosing reading material. Most persons won't mind reading a two-page magazine article printed in a small type size if it is about them. On the other hand, reading instructions on how to prepare an income tax return is uncomfortable for most of us, regardless of its type size or appearance on the page.

Some of the major factors that affect the readability of a document include: type size, font, line spacing, column width, and white space. Improper application of any of these can result in a document that is difficult to read and understand.

Serif fonts, like Times Roman, should always be used for the main body of text. Sans serif fonts, like Helvetica or Arial, can be used for headings and special notes. Because of the lack of lines at the base of each character, sans serif fonts are generally more difficult to read. For example, the text passages in this book have been printed in a serif font, while the section and chapter headings are in san serif typeface.

Font consistency can also play a role in the readability of a document. Constantly changing fonts can confuse the reader and tire the eyes. The main body of the text, except for emphasized examples, should be printed in the same serif type size and font.

Line spacing can also make a page of text easy or difficult to read. One or one and one-half line spacing is adequate for most manuals. However, double or triple line spacing should be used for draft versions of the manual to allow for notes and editing marks.

White space describes the space on a printed page that is not occupied by text or graphics. Common white space areas are the top, bottom, and side margins. Insufficient white space will tire the user's eyes. Lack of white space can also give the false impression of complexity within the text. For example, note how highlighted blocks of text, pictures, and other graphic elements interrupt the main text columns in many magazine articles. The white space surrounding these elements gives the eyes a break and allows the reader to explore graphics elements as well as text.

Using Icons

Icons are used within text to visually direct the reader or imply action or reaction. At the turn of the century, icons, such as a pointing hand, were used frequently in printed matter. If applied judiciously, icons can help the reader.

Page Headers

A page header is a short text string at the top of a printed page. The text can be a page number or the publication title or chapter name. Until recently, most O&M manuals didn't have headers because they were too difficult to add with a typewriter and were viewed as unnecessary work.

Modern word processing and desktop publishing programs have made it easier to add format elements such as headers and footers. Page headers allow users to find their location within the manual at a glance. It also allows them to thumb through the manual to find specific chapter and section titles. Page numbering and text can be formatted to accommodate two-sided printing and even-odd page numbering.

Digital Scanning

Computer technology now offers the ability to scan any printed image such as a photograph, shop drawing, line drawing, etc., into a digital file. These digital image files can be electronically placed into documents using most word processing applications. Not only can the images be integrated with text, they can also be manipulated in a variety of ways.

Graphic software can enlarge, reduce, rotate, and stretch the image. It can also alter lines and colors within the image. Any printed image can be scanned and subsequently altered. The hardware and software needed to perform image manipulation are affordable and will run on most personal computers.

There is a price to pay for digitally scanned images: hard drive space. Scanned photographs can range in size from as little as 10 kilobytes to 2 megabytes and larger. As you can imagine, it won't take too many 2-megabyte graphic files to fill up a small hard drive.

Most digital scanners will allow scanning at various resolutions. Higher resolution scans provide more clarity in the image. Lower resolution scans cause the image to appear grainy and broken. If you plan to use a lot of scanned images in the manual and want to view it on a computer, consider using medium-resolution scans. If high-resolution scans are a must, a large hard drive will be necessary.

The de facto graphic image standard for web-based applications is JPEG (Joint Pictures Entertainment Group). JPEG files are easy to manipulate with most graphics editing applications and can be inserted into web sites and web-based manuals very easily. JPEG files can also be compressed to retain quality while reducing the file size considerably.

Optical Character Recognition

Software programs are available that allow direct conversion of scanned text files into word processing and spreadsheet formats. This conversion technology is called *optical character recognition* (OCR). Any printed document can be converted into electronic format without having to retype it. Following are the steps in the OCR process:

1. The document page is scanned into an electronic image (or picture) file. At this point, it is nothing more than an electronic "picture" of the printed page. At this stage, individual text characters cannot be recognized by the computer.

2. The OCR software examines each printed character and attempts to match the character with one loaded in its reference library. It examines a printed character, such as the letter "A," and "recognizes" it as an upper case "A." The image is then replaced by the ASCII character "A." The OCR software continues down the page until it has "recognized" and replaced all the characters.

The file can then be saved as a computer word processing, spreadsheet, or database file.

Digital scanners and OCR software can be an inexpensive alternative to retyping hundreds of pages of text. OCR software isn't perfect, however. Its ability to recognize printed characters can be limited by factors such as the quality of the printed page, the types of fonts used, the alignment of text, etc. Some character recognition software programs can be "trained" to recognize certain types of printing and even the idiosyncrasies of an individual's handwriting.

The main advantage to using OCR software is the time savings in reproducing the text into electronic form. Another benefit is the ability to manipulate the converted document with most word processing and desktop publishing programs

Producing Pocket-Sized Versions of an O&M Manual

Most plant operators prefer to have O&M manual information as accessible as possible. Operators and maintenance technicians, especially entry level, appreciate having O&M manual information at their fingertips, many preferring to take it with them in the field. If you are producing your manual on a word processor and printing it with a laser printer, you can easily produce pocket-sized manuals.

Most word processors and desktop publishing programs will allow you to print in a small, multi-column format. To avoid altering the original files, make a working copy. Review your software's instructions for formatting existing text documents to a two-column or newsletter format. The pages can be printed in two-column letter size and saddled-stitch binded (center-stapled). If there are many pages of manual text, individual sections can be printed and supplied with cover sheets. New pocket versions of the manuals can be printed as they wear or information is updated in the manual. The manuals can even be printed on moisture resistant paper to extend their life.

Chapter 5

Producing and Maintaining an Online Manual

Most of us interact with computers on a daily basis. Every facet of our society uses computers to monitor and control plant processes. Documentation and training materials are produced using computers, and these electronic documents can play an important role in facility training as well as day-to-day operations.

Online O&M manuals are becoming very popular as well. Many facilities do not have current, up-to-date manuals, and new computer technologies offer an ideal alternative for creating, preserving, and updating these manuals. Online O&M manuals allow the user to easily add or modify text and drawings as equipment and operating procedures change. This versatility greatly increases the accuracy and usefulness of the O&M manual and eliminates reliance on outside organizations for its upkeep.

Since features of the facility change during final design and construction, the O&M manual must be edited and portions rewritten to reflect these changes. This critical step of updating the O&M manual is often ignored. After construction of the facility is complete and the contractors, engineers, planners, and administrators have gone, facility operators and maintenance technicians are often left with an inaccurate or incomplete O&M manual. The manual cannot be easily corrected because of the expense and difficulty of producing a new hardbound copy. Updating the diagrams and drawings and re-indexing the manual add to the difficulty of the task.

An online O&M manual can be easily created for virtually any computer system. An Apple Macintosh, IBM PC or compatibles running Microsoft Windows, UNIX based computers, or even larger systems such as IBM mainframes can be used to create and view an online manual. Facilities equipped with a local area network (LAN) make it possible for the manual to be accessed from all computers connected to the LAN. The online manual can also be used as the primary tool for integrating other plant information systems. In general, online O&M Manuals should include the following:

- Text documents regarding equipment and process operating procedures
- CADD drawings
- Scanned photographs
- Scanned images of manufacturer/vendor information
- Navigation system (menus/graphics)

Any good online O&M manual must have an efficient menu navigation system. With a good menu system, the user can select a topic from a menu list or graphic element on the screen. The online manual will then either display the information requested or jump to another menu in the manual hierarchy.

The user can create any type of menu organization. The menu's primary function is to allow the user to navigate easily through the text and graphics of the O&M manual. Menus lead to other menus, or to text or graphics files. Depending upon the menu system, the original document file can be used for viewing and editing, making translation to another file format unnecessary. The creation of the original menus and subsequent menu revisions need to be determined with input from the end-user.

Before developing the menu, a plan or *storyboard* should be created to define how the parts of the manual fit together, and what specific text and graphics files are to be used. A storyboard can be thought of as the "organizational chart" of the manual, with one menu branching to a lower menu, and so on until the desired text and graphics files are reached.

An online O&M manual is an excellent starting point in the development of a facility-wide information management system. A facility's CADD drawings, O&M text, scanned photos, and other information need to be integrated with equipment maintenance information, training programs, and other information. As the system is expanded, budget tracking and other accounting systems can be integrated to provide staff with up-to-date information.

Sharing and accessing information quickly is the key to getting more work done in less time. If most of your time is spent gathering information, you will have less time to analyze information. Online manuals can provide a reliable source of operations and maintenance information.

Web Browser-based Manuals

Web browser software such as Netscape Communicator and Microsoft Internet Explorer allow the user to link information within a document or graphic file to other information within the same file, or other files based on individual words or phrases.

For example, an online manual user might come across the word "pump" in a specific section of text. By selecting (clicking on) a word or phrase, the user is immediately redirected to another section of the manual that might include the definition of the word, a list of pumps by building at a facility, or a graphical representation of a particular pump. These referenced sections can then reference other sections, and so on.

The possibilities for text linking combinations are truly endless. The advantage to creating and using web browser-based documents is the consolidation and integration of information. The concept of documents with indexes, tables of contents, and headings is replaced by the hypertext linking features. Information within large documents can be located quickly and efficiently.

Developing Manuals Using HTML

Hypertext markup language, or HTML, is the language of the Internet and the language protocol used by web browsers. However, HTML *is not* a programming language; therefore, you do not need to be a computer programmer to build simple HTML web browser-based online manuals. You do need to become familiar with HTML itself if you plan to create online manuals. Many good books are available on the subject of HTML. Most of these texts provide basic information on HTML as well as intermediate and advanced techniques. Anyone with a working knowledge of Microsoft Windows applications can learn basic HTML.

Creating and Editing HTML Files

HTML was created as a simple markup language. Because it uses only the ASCII set of characters, a simple text editor such as Microsoft Notepad can be used to edit HTML files. A number of very good HTML text editors are available that provide access to markup language tags and other components. They also allow instant viewing of the HTML page files so you can check your work as you go. These editing programs are very affordable and some can even be downloaded directly from the Internet. Following is a discussion of several online manual components and how HTML can be used to incorporate them.

Manual Text

The greatest cost in developing any O&M manual, including online manuals, is the writing of the text documents. For an online manual, the author simply writes the manual text using a popular word processor. These files are edited, then checked and reviewed by facility staff.

Most popular word processing programs allow the conversion of text directly to HTML format for viewing in a Web browser. These word processors not only convert text, but also images for placement in the HTML file. Be aware, however, that text formatting is often altered when HTML is applied. Also, tables and image size and placement can vary considerably from the original document. Many consider these limitations necessary in order to maintain Web browser viewing flexibility. However, others feel that it unduly alters the original document. Should it become necessary to always display a document in its original format, another Web browser displayable format such as Adobe portable document format (PDF) should be used. PDF files are discussed later in this chapter.

Text files can also be indexed to facilitate locating specific information within the manual. After indexing, words, phrases, and even entire files can be located in just a few seconds. Many web search engines are available that will index and search HTML files. Most books on HTML discuss the use of search engines.

CADD Drawings

Facility and equipment drawings are one of the most valuable tools for staff. With accurate and accessible drawings, important information regarding the equipment can be located quickly and efficiently. An online O&M manual can make CADD drawings very accessible to the user without taking up valuable office space.

Normally, a facility will have two types of drawings: hard copy and/or digital files. Most facilities will have a combination of the two. Obviously, when integrating drawings with an online manual system, drawing files in digital format are much more valuable. Hard copy drawings can be incorporated into the online manual using digital scanners.

A number of web browser-based CADD drawing viewers are available. Many of these viewers work directly within the web browser. These programs are known as *plugins*. Once the file extension has been associated with the browser plugin, (i.e., DWG, DGN, DXF, etc.), any CADD drawing file can be viewed directly within the browser. This means that the end-users need not have a high-end CADD program installed on their computer in order to view the CADD files. Many web browser plugins also allow drawing markup and redlining. For example, AutoDesk, the makers of AutoCAD, have a freely downloadable AutoCAD viewing application called VoloView Express. You can download VoloView Express at *www.autodesk.com.*

Converting printed drawing files to CADD files can be very expensive. Many times all that is needed is a good quality scan of the drawing that can be viewed in a web browser. Programs like Adobe Acrobat provide an excellent platform for this. Full-size or half-size CADD drawings can be quickly and affordably scanned and viewed in any web browser. However, it must be noted that the better the original hard copy quality, the better the displayed image.

Scanned Images

Photographs and figures can also be included in online manuals. These items can be scanned and saved into the appropriate graphics file format. When including these images into a web browser-based O&M manual, the files should be saved in JPEG or GIF format. These two graphics file formats have been adopted as de facto standards on the World Wide Web. The general rule is to use JPEG format when scanning pictures or items with a large number of colors and to use GIF format for items having 256 colors or less.

Digital Cameras

Digital cameras allow you to take and save photographs directly to digital format. Most digital cameras will not produce quality normally found with film cameras, but can provide graphic quality that is acceptable for most online manuals. Two advantages of using a digital camera are that there are no film processing costs, and the photograph need not be scanned into digital format.

PDF Documents

Adobe Systems Acrobat software allows the creation and viewing of portable document format (PDF) files. PDF has become the de facto standard for scanned searchable documents on the Internet. Most federal and state governmental agencies have adopted the PDF standard for downloadable documents. Adobe Reader can be freely downloaded from the Adobe website *www.adobe.com*. Therefore, once a document has been converted to PDF format, it can be viewed by anyone using Adobe Reader. Adobe Reader can also be used as a web browser plugin, allowing PDF documents to be viewed directly in the web browser.

What are the advantages to saving a document in PDF format? When converting a text document to HTML, much of the formatting can be inadvertently lost or altered. When a document is converted to PDF format, the exact look and feel of the original document is maintained. Any graphics such as tables, graphs, photographs, and logos are also presented in their original size and position in the document. Any printable document can be converted to PDF format. PDF files are often much larger than their HTML counterparts, but in cases where the original format of the document must be maintained, it is definitely the format of choice.

PDF files can also be indexed and searched. Keywords can be added to PDF files to facilitate searching and separate indexes can be created for different file types. Adobe Acrobat also provides for automatic scheduled indexing of frequently changing documents.

Video

Analog videotapes can be converted to standard digital file formats such as MPEG and AVI. Once the videotapes have been converted, they can be edited with readily available software. You can also obtain video capture cards that can be installed in most desktop computers. Once installed, you can plug a VCR or video camera into the capture card and convert the tape to digital format. Most video capture cards come with digital video editing software. Video capture cards can cost as little as $600 and as much as $2,000, depending upon your needs.

Once converted to digital format, video files can be played on most Microsoft Windows and Macintosh computers. Most popular web browsers also support the MPEG and AVI file formats.

Index

Date Due

11/27/03 IL440942		
IL17519067		4.26.17

BRODART, CO.   Cat. No. 23-233-003   Printed in U.S.A.